Compendium of Potato Diseases

W. J. Hooker, Editor

In cooperation with:
The Potato Association of America
The International Potato Center, Lima, Peru
Department of Botany and Plant Pathology, Michigan State University
Michigan State Foundation

APS PRESS

The American Phytopathological Society

Covers: Plants of the indigenous South American triploid cultivar
Huayro (*Solanum* $\times$ *chaucha,* 2n = 3x = 36), in flower.
W. J. Hooker

Library of Congress Catalog Card Number: 80-85459
International Standard Book Number: 0-89054-027-6

©1981 by The American Phytopathological Society
Second Printing, 1983
Third Printing, 1986
Fourth Printing, 1990

Printed in the United States of America

The American Phytopathological Society
3340 Pilot Knob Road
St. Paul, Minnesota 55121, USA

In memory of my respected colleague

Joseph E. Huguelet
1931–1977

Preface

This compendium, a compilation of information describing diseases and disorders of the potato, is intended to be applicable throughout the world, including the tropics.

For some time a need has existed for a source of information concerning potato diseases and production problems associated with disease prevention. Because diseases frequently limit successful production, and production is an expensive venture, concise, up-to-date information is required.

Accounts of diseases were prepared by individuals chosen for their experience and knowledge of the subject. Literature on the important diseases is extensive, and thorough understanding of a single major disease has involved many investigations. References were selected to provide recent information and access to previously published literature.

The compendium results from the combined efforts of many individuals and several organizations. Members of the advisory committee provided guidance in coordination and initial planning as well as in critical and constructive manuscript review. Many others have generously given their time in review of parts of the manuscript.

Qualified persons prepared descriptions of individual diseases. These and other persons and certain organizations have graciously loaned photographs. Financial assistance from persons and organizations has permitted inclusion of pertinent colored plates. Unless otherwise indicated in the caption, photographs in the text are from my files.

The continual encouragement and cooperation of the Potato Association of America and its members have been most gratifying, encouraging, and constructive.

Without assistance from Michigan State University (MSU), the International Potato Center, and the Michigan Foundation, this work could not have been completed. All have provided grants to enable manuscript preparation. The International Potato Center and MSU have also provided research facilities. Cooperation and professional capabilities of the library staff and personnel of the Graphic Arts Center (MSU) is greatly appreciated.

The continued enthusiastic cooperation of Dr. Teresa Icochea in critically reviewing this manuscript is sincerely appreciated.

The stenographic expertise, attention to detail, and patience of Suzanne J. Weise, Monica Stenning, and Elaine Creech in manuscript preparation have been invaluable.

To Frances, my wife, for her patience, encouragement, and willingness to forego other interests so that this compendium might be completed, I am grateful.

W. J. Hooker

DISCLAIMER

Products and practices included in the control paragraph of each disease are those reported to be effective. Permitted use of chemicals varies among the nations producing and consuming potatoes. The international nature of this publication precludes limiting preventive measures to those accepted by particular governmental control agencies. Authors, sponsors, and organizations under whose auspices the compendium was prepared assume no responsibility for effectiveness of chemicals nor for authorizing or recommending use of chemicals or other preventive measures mentioned in this publication. Recommendations for chemical selection, dosage, and method and time of application should be made only by authorized and informed personnel of the responsible governmental agency where potatoes are grown or marketed, and these instructions should be rigidly followed.

Acknowledgments

Planning Committee

W. J. Hooker, Coordinator, Michigan State University, East Lansing.
Present address: International Potato Center, Lima, Peru

Edward R. French, International Potato Center, Lima, Peru

Robert W. Goth, USDA Plant Genetics and Germplasm Institute,
Beltsville, MD

Monty D. Harrison, Colorado State University, Fort Collins

*J. E. Huguelet, North Dakota State University, Fargo

Arthur Kelman, University of Wisconsin, Madison

W. F. Mai, Cornell University, Ithaca, NY

Frank Manzer, University of Maine, Orono

James Munro, 561 Dickinson Ave., Ottawa, Ontario, Canada

L. W. Nielsen, North Carolina State University, Raleigh

Dick Peters, Agricultural University, Wageningen, The Netherlands

*Otto E. Schultz, Cornell University, Ithaca, NY

N. S. Wright, Canada Department of Agriculture Research Station,
Vancouver, B.C.

*Deceased

Contributors of Manuscripts or Photographs

Agrios, G. N., University of Massachusetts, Amherst

ASARCO Inc., Department of Environmental Sciences, Salt Lake
City, UT

Bagnall, R. H., Agriculture Canada Research Station, Fredericton,
N.B.

Beemster, A. B. R., Research Institute for Plant Protection,
Wageningen, The Netherlands

Bennett, C. W., USDA, Salinas, CA

Berger, K. C., Berger and Associates, Hartland, WI

Bhattacharyya, S. K., Central Potato Research Institute, Simla
(H.P.) India

Boawn, L. C., Washington State University, Pullman

Booth, R., International Potato Center, Lima, Peru

Boyle, J. S., Pennsylvania State University, University Park

Brodie, B. B., Cornell University, Ithaca, NY

Bryan, J., International Potato Center, Lima, Peru

Burton, W. G., East Malling Research Station, Maidstone, Kent,
England

Busch, L. V., University of Guelph, Guelph, Ont., Canada

Butzonitch, I. P., Instituto Nacional de Technologia Agropecuaria,
Balcarce, Argentina

Calderoni, A. V., Instituto Nacional de Technologia Agropecuaria,
Balcarce, Argentina

Costa, A. S., Instituto Agronomico de Estado de Sao Paulo, Sao
Paulo, Brazil

Darling, H., University of Wisconsin, Madison

Davis, J. R., University of Idaho, Aberdeen

de Abad, G., International Potato Center, Lima, Peru

de Bokx, J. A., Research Institute for Plant Protection, Wageningen,
The Netherlands

de Icochea, T. A., Universidad Nacional Agraria, Lima, Peru

de Zoeten, G., University of Wisconsin, Madison

Dwivedi, R., Central Potato Research Institute, Simla (H.P.) India

Dyson, P. W., Macaulay Institute of Soil Research, Craigiebuckler,
Aberdeen, Scotland

Easton, G. D., Irrigated Agriculture Research and Extension Center,
Prosser, WA

Ellis, M. B., Commonwealth Mycological Institute, Kew, Surrey,
England

Ewing, E. E., Cornell University, Ithaca, NY

Fernow, K. H., 1228 Ellis Hollow Road, Ithaca, NY

Frank, J. A., Pennsylvania State University, University Park

French, E. R., International Potato Center, Lima, Peru

Frey, F., International Potato Center, Lima, Peru

Fribourg, C. E., Universidad Nacional Agraria, Lima, Peru

Gausman, H. W., U.S. Fruit and Vegetable Soil and Water Research,
Weslaco, TX

Genereux, H., Ministere de l'Agriculture de Canada, Sainte-Anne-de-
la-Pocatiere, Quebec, Canada

Goth, R. W., USDA Plant Genetics and Germplasm Institute,
Beltsville, MD

Hampson, M. C., Agriculture Canada Research Station, St. John's,
Newfoundland

Harrison, B. D., Scottish Horticultural Research Institute,
Invergowrie, Dundee, Scotland

Harrison, M. B., Cornell Nematology Laboratory, Farmingdale, NY

Harrison, M. D., Colorado State University, Fort Collins

Hide, G. A., Rothamsted Experiment Station, Harpenden, Herts,
England

Hiruki, C., University of Alberta, Edmonton

Hooker, W. J., Michigan State University, E. Lansing

Huguelet, J. E., North Dakota State University, Fargo

Huttinga, H., Research Institute for Plant Protection, Wageningen,
The Netherlands

International Minerals and Chemical Corp., Libertyville, IL

Iritani, W. M., Washington State University, Pullman

Jatala, P., International Potato Center, Lima, Peru

Jones, E. D., Cornell University, Ithaca, NY

Jones, R. A. C., International Potato Center, Lima, Peru

Keller, E. R., Institut für Pflanzenbau der Eidgenössischen
Technischen Hochschule, Zürich, Switzerland

Kelman, A., University of Wisconsin, Madison

Kitajima, E. W., Instituto Agronomico de Estado de Sao Paulo, Sao
Paulo, Brazil

Larson, R. H., University of Wisconsin, Madison

Laughlin, W. M., USDA, Palmer, AK

Lawrence, C. H., Agriculture Canada Research Station, Fredericton,
N.B.

Leggett, G. E., Utah State University, Logan

Lennard, J. H., Edinburgh School of Agriculture, Edinburgh,
Scotland

Letal, J. R., Regional Crops Laboratory, Olds, Alberta, Canada

Logan, C., Department of Agriculture, Belfast, N. Ireland

MacGillivray, M. E., Agriculture Canada Research Station,
Fredericton, N.B.

Mai, W. F., Cornell University, Ithaca, NY

Manzer, F. E., University of Maine, Orono

Martin, C., International Potato Center, Lima, Peru

McIntyre, G. A., Colorado State University, Ft. Collins

McKenzie, A. R., Agriculture Canada Research Station, Fredericton,
N.B.

Munro, J., 561 Dickinson Ave., Ottawa, Ontario, Canada

Nelson, E., Colorado State University, Ft. Collins

Nelson, P. E., Pennsylvania State University, University Park

Nielsen, L. W., North Carolina State University, Raleigh

Page, O. T., International Potato Center, Lima, Peru

Peters, D., Agricultural University, Wageningen, The Netherlands

Raine, J., Canada Department of Agriculture, Research Station,
Vancouver, B.C.

Rich, A. E., University of New Hampshire, Durham

Rowe, R. C., Ohio Agricultural Research and Development Center, Wooster
Salazar, L. F., International Potato Center, Lima, Peru
Salzmann, R., Eidgenössischen Forschungsanstalt für landwirtschaftlichen Pflanzenbau, Zürich-Reckenholz, Switzerland
Schultz, O. E., Cornell University, Ithaca, NY
Sinden, S. L., USDA Plant Genetics and Germplasm Institute, Beltsville, MD
Singh, R. P., Agriculture Canada Research Station, Fredericton, N.B.
Slack, S. A., University of Wisconsin, Madison
Sparks, W. C., University of Idaho, Aberdeen
Stace-Smith, R., Agriculture Canada Research Station, Vancouver, B.C.
Thurston, H. D., Cornell University, Ithaca, NY
Torres, H., International Potato Center, Lima, Peru
Turkensteen, L. J., International Potato Center, Lima, Peru
Ulrich, A., University of California, Berkeley
Untiveros, D., International Potato Center, Lima, Peru
Valenta, V., Slovak Academy of Sciences, Bratislava, Czechoslovakia
Weingartner, D., University of Florida, Hastings
Wright, N. S., Agriculture Canada Research Station, Vancouver, B.C.
Zachmann, R., International Potato Center, Lima, Peru

Financial Sponsors

Agan's Aviation, Inc., Munger, MI
American Hoechst Corp., Somerville, NJ
BASF Aktiengesellschaft, Ludwigshafen/Rhein, Germany
British Columbia Coast Vegetable Co-operative Assoc., Richmond, B.C., Canada
Blue Mountain Potato Growers Assoc., Hermiston, OR
Brown, Miles, Elmira, MI
Chevron Chemical Co., Ortho Division, San Francisco, CA
Christensen, Ferris H., Edmore, MI
CIBA-Geigy Corp., Greensboro, NC
Diamond Shamrock Corp., Cleveland, OH
E. I. du Pont de Nemours and Co., Wilmington, DE

FMC Corp., Agricultural Chemical Group, Philadelphia, PA
Fridell, E. W., Plant Quarantine Division, Barrie, Ont., Canada
Gamez Produce Co., Hereford, TX
Great Lakes Chemical Corp., West Lafayette, IN
Greenberg Farms, Art, Grand Forks, ND
Growers Service Corp., Lansing, MI
Imperial Chemicals Industries United States, Inc., Wilmington, DE
Idaho Crop Improvement Assoc., Boise, ID
Krueger, Paul, Jr., Potato Seed, Hawks, MI
Lamb-Weston, Division of AMFAC Foods, Inc., Portland, OR
Lennard, Wayne J., & Sons, Inc., Samaria, MI
Malheur County Potato Growers Assoc., Ontario, OR
Merck and Co., Inc., Agricultural Chemicals Development, Rahway, NJ
Merck Sharp and Dohme International, Rahway, NJ
Michigan Crop Improvement Assoc., Lansing
Michigan Potato Industry Commission, Lansing
Mobay Chemical Corp., Chemagro Division, Kansas City, MO
Montcalm County Board of Commissioners, Stanton, MI
New York Seed Improvement Cooperative, Inc., Ithaca
North Dakota Certified Seed Potato Growers Assoc., Inc., Fargo
North Dakota State Seed Department, Fargo
Olin Agricultural Products Department, Little Rock, AR
Oregon Potato Commission, Salem
Ore-Ida Foods, Inc., Boise, ID
Perham Potato Farm, Fergus Falls, MN
Richardson, Tuck, Lakeland Farm, Gilbert, MN
Rohm and Haas Co., Philadelphia, PA
Sackett Ranch, Stanton, MI
Schalk, Allen and Paul, Rogers City, MI
Schering AG, D-1000 Berlin 65, Germany
Snake River Chemicals, Inc., Caldwell, ID
Stauffer Chemical Co., Mountain View, CA
Thompson-Hayward Chemical Co., Kansas City, KS
Union Carbide Corp., Agricultural Products Division, Salinas, CA
Velsicol Chemical Corp., Chicago, IL
Wisconsin Seed Improvement Assoc., Madison

Contents

Introduction

- 1 **Potato Disease**
- 1 **The Potato**
- 1 Importance
- 2 Cultivated Types
- 2 The Plant
- 5 **Oxygen-Temperature Relationships**

Part I. Disease in the Absence of Infectious Pathogens

- 7 **Genetic Abnormalities**
- 8 **Adverse Environment**
- 8 Oxygen Deficit
- 8 Low Temperature Tuber Injury
- 9 Low Temperature Foliage Injury
- 10 Blackheart
- 10 High Temperature Field Injury
- 11 Internal Heat Necrosis
- 12 Second Growth and Jelly End Rot
- 13 Hollow Heart
- 14 Surface Abrasions
- 15 Blackspot
- 16 Tuber Greening and Sunscald
- 17 Internal Sprouting
- 17 Secondary Tubers
- 17 Coiled Sprout
- 18 Hair Sprout
- 18 Nonvirus Leafroll
- 19 Hail Injury
- 19 Wind Injury
- 19 Lightning Injury
- 20 Air Pollution: Photochemical Oxidants
- 21 Air Pollution: Sulfur Oxides
- 21 Chemical Injury
- 22 Stem-End Browning
- 22 **Nutrient Imbalance**
- 22 Nitrogen
- 23 Phosphorus
- 23 Potassium
- 23 Calcium
- 24 Magnesium
- 24 Sulfur
- 24 Aluminum
- 24 Boron
- 25 Zinc
- 26 Manganese

Part II. Disease in the Presence of Infectious Pathogens

- 27 **Bacteria**
- 27 Blackleg, Bacterial Soft Rot
- 29 Brown Rot
- 31 Ring Rot
- 32 Pink Eye
- 33 Bacteria in Potatoes that Appear Healthy
- 33 Common Scab
- 35 **Fungi**
- 35 Powdery Scab
- 36 Wart
- 37 Skin Spot
- 38 Leak
- 39 Pink Rot
- 40 Late Blight
- 42 Powdery Mildew
- 43 Early Blight
- 44 *Alternaria alternata*
- 46 *Pleospora herbarum*
- 46 Ulocladium Blight
- 46 *Stemphylium consortiale*
- 46 Septoria Leaf Spot
- 47 Cercospora Leaf Blotches
- 47 Phoma Leaf Spot
- 48 Choanephora Blight
- 48 Gray Mold
- 48 White mold
- 50 Stem Rot
- 51 Rosellinia Black Rot
- 52 Rhizopus Soft Rot
- 52 Rhizoctonia Canker (Black Scurf)
- 54 Violet Root Rot
- 54 Silver Scurf
- 55 Black Dot
- 56 Charcoal Rot
- 57 Gangrene
- 58 Fusarium Dry Rots
- 60 Fusarium Wilts
- 62 Verticillium Wilt
- 63 Thecaphora Smut
- 65 Common Rust
- 65 Deforming Rust
- 66 Miscellaneous Diseases
- 66 Mycorrhizal Fungi
- 66 Principles of Foliage Fungicide Application
- 67 Tuber Seed Treatment
- 68 **Viruses**
- 68 Potato Leafroll Virus
- 70 Potato Virus Y
- 71 Potato Virus A
- 72 Potato Virus X
- 74 Potato Virus M
- 75 Potato Virus S
- 77 Potato Virus T
- 77 Andean Potato Mottle Virus
- 78 Andean Potato Latent Virus
- 79 Cucumber Mosaic Virus
- 79 Tobacco Mosaic Virus
- 79 Potato Mop-Top Virus

80 Tobacco Rattle Virus
82 Potato Yellow Dwarf Virus
82 Alfalfa Mosaic Virus
84 Potato Aucuba Mosaic Virus
84 Tobacco Ringspot Virus
85 Tomato Black Ring Virus
86 Potato Yellow Vein Virus
86 Tobacco Necrosis Virus
87 Deforming Mosaic
87 Tomato Spotted Wilt Virus
89 Potato Spindle Tuber Viroid
90 Sugar Beet Curly Top Virus
91 **Mycoplasmas**
92 Aster Yellows and Stolbur
92 Witches' Broom
93 **Insect Toxins**
93 Psyllid Yellows

93 **Nematodes**
94 Potato Cyst Nematodes
97 Root-Knot Nematodes
98 False Root-Knot Nematodes
99 Lesion Nematodes
100 Potato Rot Nematodes
101 Stubby-Root Nematodes
101 Nematicides
101 **Aphids**
103 **Seed Potato Certification**

107 **Key to Disease**
111 **Equivalent Names of Potato Diseases**
Color Plates (following page 54)
117 **Glossary**
123 **Index**

Introduction

Potato Disease

A potato disease is an interaction between a host (the potato) and a pathogen (bacterium, fungus, virus, mycoplasma, nematode, or adverse environment) that impairs productivity or usefulness of the crop. Frequently, adverse environmental effects are sufficient to initiate disease in the absence of an infectious entity. The host-pathogen interaction is influenced by environment acting on either the potato or the pathogen or on both and is determined by the genetic capabilities of 1) the potato in being either susceptible or resistant and 2) the pathogen in being pathogenic (virulent) or nonpathogenic (avirulent).

Furthermore, disease or adverse environment in one portion of the potato life cycle may severely limit effectiveness of production or quality at a later date. For example, field problems frequently become storage problems, which may later limit either market quality or seed performance and, ultimately, yielding ability.

The value of any crop determines the extent to which control measures may be justified. Relatively speaking, the potato is a high-value crop with complex production, storage, and utilization problems, and therefore relatively elaborate prevention practices are appropriate. Correct diagnosis and identification of disease is of paramount importance for initiation of appropriate control and prevention measures.

General References

BLODGETT, E. C., and A. E. RICH. 1949. Potato tuber diseases, defects, and insect injuries in the Pacific Northwest. Wash. Agric. Exp. Stn. Pop. Bull. No. 195. 116 pp.

CALDERONI, A. V. 1978. Enfermedades de la papa y su control. Editorial Hemesferia Sur S. A., Buenos Aires. 143 pp.

FRENCH, E. R., H. TORRES, T. A. de ICOCHEA, L. SALAZAR, C. FRIBOURG, E. N. FERNANDEZ, A. MARTIN, J. FRANCO, M. M. de SCURRAH, I. A. HERRERA, C. VISE, L. LAZO, and O. A. HIDALGO. 1972. Enfermedades de la Papa en el Perú. Bol. Tecn. No. 77 Est. Exp. Agric. La Molina. 36 pp.

HODGSON, W. A., D. D. POND, and J. MUNRO. 1974. Diseases and pests of potatoes. Canada Dept. Agric. Publ. 1492. 69 pp.

KELLER, E. R., and A. ZAH. 1969. Dictionary of Technical Terms Relating to the Potato (in English, German, and French). Eur. Assoc. Pot. Res. Juris-Druck, Zurich. 111 pp.

McKAY, R. 1955. Potato Diseases. Irish Potato Marketing Co., Ltd., Dublin. 126 pp.

O'BRIEN, M. J., and A. E. RICH. 1976. Potato Diseases. U.S. Dept. Agric., Agric. Res. Serv., Agriculture Handbook No. 474. 79 pp.

SALZMANN, R., and E. R. KELLER. 1969. Krankheiten und Schadlinge der Kartoffel. Verbandsdruckerei. AG, Bern. 150 pp.

SCHICK, R., and M. KLINKOWSKI. 1961. Die Kartoffel, ein Handbuch. Vol. I. 1,007 pp.; Vol. II, 2,112 pp. Veb Deutscher Landwirtschaftsverlag, Berlin.

SMITH, W. L., Jr., and J. B. WILSON. 1978. Market diseases of potatoes. U.S. Dept. Agric., Agric. Handbook No. 479. 99 pp.

WHITEHEAD, W. T., P. McINTOSH, and W. M. FINDLAY. 1953. The potato in health and disease, 3rd ed. Oliver and Boyd, London. 774 pp.

(Prepared by W. J. Hooker)

The Potato

Importance

The potato is the most important dicotyledonous source of human food. It ranks as the fifth major food crop of the world, exceeded only by the grasses—wheat, rice, maize (corn), and barley. In North America, dry matter production of potatoes per unit of land area exceeds that of wheat, barley, and maize by factors of 3.04, 2.68, and 1.12, respectively. Yields of protein per unit of land area exceed those of wheat, rice, and maize by factors of 2.02, 1.33, and 1.20, respectively.

Because of increasing yields per unit area of land, total potato production has been increasing even though the area of land planted to potatoes is decreasing. Yields in northern Europe and North America (1970–1973) generally ranged from 20 to over 35 metric ton/ha (178–311 cwt/A) and were somewhat lower in the warmer areas of Europe. The percentage of arable land in potato production ranges from less than 1% in Canada and the United States to 18% or more in the Netherlands and Poland. The USSR, China, and Poland lead in area of land in potato production.

In the tropics (between 30° north and south latitudes) yields are below 13 ton/ha (116 cwt/A), more commonly below 10 tons (89 cwt/A), and the percentage of arable land in potatoes is low except in Peru (13%). However, potato production is increasing as measured by land area in potato production, yield per hectare, and total production.

Most potatoes are used for human consumption. In the tropics potatoes are often available only in certain seasons because of storage problems. Approximately 50% of European potato production is used as stock feed, with perhaps 25% of table stock potatoes being diverted to stock feed because of defects.

The potato is characteristically a crop of the cool, temperate regions or of elevations of approximately 2,000 m (6,560 ft) or more in the tropics. It requires cool nights and well drained soil with adequate moisture and does not produce well in low altitude, warm, tropical environments. Certain types of South American potato have considerable tolerance to warm temperatures and to temperatures a few degrees below freezing; clones tolerant to both extremes are being sought.

The potato is a native of the Andean highlands of South America, where it has served as a staple of the diet of Andeans for centuries or millennia. Widely diverse types have been selected. Dehydrated tubers have been preserved since antiquity as chuño, a product of various types obtained by freezing tubers, pressing liquid from them after thawing, and then freezing and drying them. Today tubers are boiled and dried for preservation.

The potato was introduced into Spain sometime before 1573, when it was first mentioned as a food source. Its use in England was first reported in herbals in 1596. From northern Europe it was first reported in herbals in 1596. From northern Europe it was returned to North America in 1719 and grown in the colonies.

Cultivated Types

Cultivated potatoes, consisting of a number of species or species hybrids, belong to the Solanaceae, section Tuberarium, which contains approximately 150 tuber-bearing species. In the Andean highlands, the distinction between cultivated and wild varieties has little relevance to indigenous populations.

The most common potato is the tetraploid ($2n = 4x = 48$ chromosomes), *Solanum tuberosum* L., which may be divided into the completely cross-fertile subgroups Tuberosum and Andigena or into subspecies *tuberosum* and *andigena*. Survival of these in the wild occurs only in exceptional cases. Their survival and extensive dispersal have resulted from human selection. Andigena is the most widely grown in South America. It has deep eyes, is often pigmented, and produces tubers in days of short length. The Tuberosum type grown in northern Europe and North America tends to need a long day for effective tuberization. These two types are not completely distinct, and obtaining the Tuberosum type by selection from Andigena is possible.

Another classification also places certain diploids and triploids within *S. tuberosum*. Diploid cultivated potatoes ($2n = 2x = 24$) belong to two main groups, *S. stenotomum*, with a period of tuber dormancy, and *S. phureja*, without a well-defined dormant period. *S. stenotomum* is regarded as the ancestral type, giving rise to Andigena types through chromosome doubling.

Cultivated triploids ($2n = 3x = 36$) in group *S. × chaucha* are possibly naturally occurring hybrids from crosses between Andigena and Stenotomum or Phureja. Another triploid species, *S. × juzepczukii*, is highly frost tolerant and may have originated from natural hybridization between the wild species *S. acaule* (a tetraploid) and the diploid *S. stenotomum*.

A pentaploid, *S. × curtilobum* ($2n = 5x = 60$), believed to have originated from natural hybridization of *S. acaule* and *S. andigena*, is grown in the high Andes because of its frost tolerance. It is bitter tasting but useful for making chuño.

The hexaploid *S. demissum* ($2n = 6x - 72$) has been used as a source for cultivars resistant to late blight.

Perpetuation of the many genetic differences is attributable to asexual propagation through tubers. Variability existing within the potato groups is believed to have originated through: 1) hybridization between diverse types, 2) chromosome doubling, 3) mutations in germ plasm, and 4) mutations in vegetative tissue and perpetuation as chimaeras. Vegetatively propagated clone lines, assumed to be genetically stable, as are presently accepted cultivars, are capable of further variation and change through somatic mutation.

The potato may be propagated from true seed, which is common in genetic studies and in potato breeding programs. In certain parts of Ecuador and Colombia true seed is planted for commercial crop production. The Peoples Republic of China grows over 10,000 ha from true seed to avoid virus spread and long distance transport of seed tubers. At the International Potato Center, transplants from true seed produce homogeneous families that yield an average of more than 1 kg per plant. These practices permit full use of tubers as food and avoid diseases carried through seed tubers.

Commercial production of most potatoes is primarily through vegetative propagation by means of lateral buds formed on the tuber, a modified stem. Through such vegetative propagation, many diseases are transmitted from generation to generation.

Although disease affects each of the types listed above, much of potato pathology has been done with *S. tuberosum* ssp. *tuberosum* in northern Europe and North America. Increased importance of the crop in the tropics and subtropics has occasioned unprecedented work during the past half century on tropical diseases and evaluation of wild and cultivated genotypes as sources of disease resistance.

The Plant

The potato, *S. tuberosum* ssp. *tuberosum* and ssp. *andigena*, is an annual, herbaceous dicotyledonous plant with potential perennial capacity because of reproduction through tubers.

Flowers

Flowers are five-parted of various colors with single style and stigma and two-loculed ovary. Pollen is typically windborne. Self-fertilization is natural; cross-fertilization is relatively rare, and when it occurs, insects are probably involved. Diploids are self-incompatible with very few exceptions.

Fruits

Fruits are round to oval (1–3 cm or more in diameter), green, yellowish green or brown, and red to violet when ripe. Fruits are two-celled, with up to 200–300 seeds. Because of several sterility factors, seeds may be absent even though fruits are formed.

Vegetative Structures

Plants from true seed are typically seedling plants with primary tap root, hypocotyl, cotyledons, and epicotyl, from which a stem and foliage develop. In contrast, the commerical potato plant contains one or more lateral branches, each arising from a bud on the "seed tuber," and the roots are adventitious (Fig. 1). The "seed" in commerical production is an asexual propagative organ and not comparable to the sexually derived true seed.

Stems

These are usually green, but can be red to purple, angular, and nonwoody. Late in the season the lower portions may be relatively woody, however. Leaves are pinnately compound, although early leaves of seedlings and the first leaves of plants grown from tubers may be simple. Leaf types differ widely among the many species and cultivars. Stomata are more numerous on the lower leaf surfaces, and hairs of various types are present on aboveground parts. Secondary branches are common, arising from axillary leaf buds. Leaves on the underground stem are small and scalelike, and stolons arise from these axillary buds. Stolons and tubers are modified adventitious stems.

Roots and stolons develop from the underground stem between the seed tuber and the soil surface. Thus, the vegetative propagative unit (the seed tuber or a portion of it, the seed piece) should be planted sufficiently deep to permit adequate root and stolon formation.

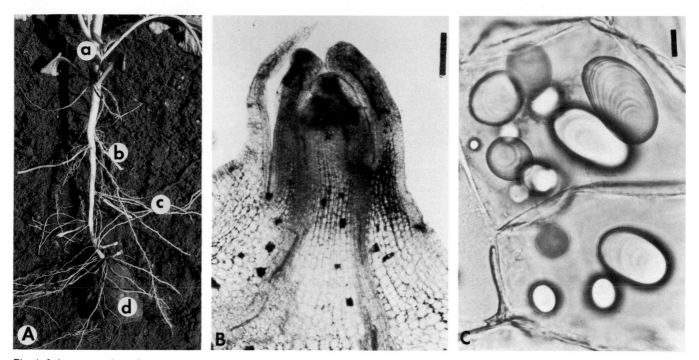

Fig. 1. **A,** Lower portion of young potato plant showing: **a,** stem; **b,** stolons; **c,** roots; and **d,** seed tuber. **B,** Bud on stolon tip (bar represents 100 µm). **C,** Starch grains within cells of a potato tuber (bar represents 10 µm), showing characteristic refraction patterns within the grain.

Tubers

The tuber (Fig. 2) is formed at the tip of the stolon (rhizome) as a lateral proliferation of storage tissue resulting from rapid cell division and enlargement. Enlargement approximates a 64-fold cell volume increase.

The stolon usually breaks off close to the tuber during harvest or dies with the plant on maturity and is evident either as a short stub or small scar.

In stems, stolons, and tubers, vascular tissue initially forms as bicollateral bundles with groups of thin-walled phloem cells outside of the xylem (outer phloem) and toward the center and inside the xylem (inner phloem). As the stolon enlarges to form the tuber, parenchyma developing within the bundles tends to split the separate groups, and the vascular ring becomes spread out. New groups of phloem including sieve tubes, companion cells, and conducting parenchyma elements are formed as the tuber enlarges. Carbohydrates are stored within storage parenchyma cells of pith and cortex in the form of starch granules with characteristic markings.

Tuber constituents vary with cultivar and growing conditions. Estimated amounts of constituents may also reflect differences in methods of chemical analysis. Ranges in the whole fresh tuber are: water, 63–87%; carbohydrates, 13–30% (including a fiber content of 0.17–3.48%); protein, 0.7–4.6%; fat, 0.02–0.96%; and ash, 0.44–1.9%. Additional constituents include sugars, nonstarchy polysaccharides, enzymes, ascorbic acid and other vitamins, phenolic substances, and nucleic acids.

The tuber surface permits or excludes entrance of pathogens, regulates rate of gas exchange or water loss, and protects against mechanical damage. The surface is not fixed and static but will maintain and regenerate itself through wound healing reactions which influence disease incidence and severity, preservation in storage, and seed germinability and performance.

The epidermis exists for only a short time on the youngest tubers of approximately 1 cm or less in diameter. Stomata are scattered in the epidermis and permit gas exchange. A short-lived periderm is derived from the epidermis and is soon replaced by a more permanent periderm or cork layer arising from meristematic cambium cells below the epidermis. This periderm in mature tubers is composed of 6–10 layers of brick-like, thin-walled cells, one on top of the other, without intercellular spaces and with suberized cell walls. Periderm

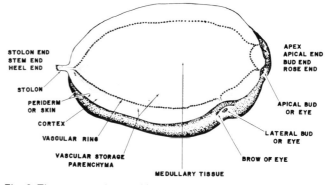

Fig. 2. The potato tuber and its parts.

characteristics vary considerably with cultivar (Fig. 3).

Wound healing (Fig. 3) develops under cut, bruised, or torn surfaces. Suberin forms within 3–5 days in walls of living cells under the wound. A cork cambium layer developing under the suberized cells gives rise to a wound periderm. The promptness with which wound healing develops is dependent upon environment (temperature, humidity, and aeration) and the physiology of the tuber.

Cut tuber surfaces exposed to drying air may seem to have a tough, resistant covering. Drying kills living surface cells and interferes with normal wound-healing activity. Dried surfaces do not exclude pathogens nor prevent dehydration and should not be confused with wound healing.

Rates of wound healing, including both suberization and periderm development, increase approximately threefold between 5 and 10°C and again threefold between 10 and 20°C. Oxygen supply less than that of the atmosphere and carbon dioxide greater than that of the atomosphere progressively inhibit wound healing. Wound healing is most rapid between 80 and 100% rh; however, the presence of free water on the surface that excludes oxygen is detrimental. Wound healing is more rapid in stored, recently harvested tubers, and as the tubers age, ability to heal wounds gradually diminishes. Also, periderm development in the cortical areas is more rapid than in the medullary region. Irradiation by sunlight or by ionizing gamma

3

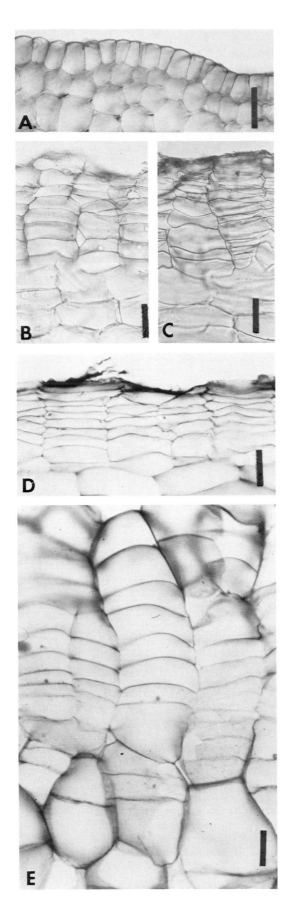

rays impairs wound-healing processes. Certain pesticide chemicals on the cut tuber surface, such as some seed treatments, may impair the effectiveness of wound healing, whereas certain phenolic compounds enhance the wound-healing process and wound periderm formation. Under favorable environmental conditions, suberin is demonstrable within 24 hr and periderm within two to five days. Infection by some wound pathogens is greatly reduced by the rapid formation of suberin and periderm under wounds.

Lenticels (Fig. 4) are formed under the stomata in the epidermis of stems as well as of tubers. A loose mass of thin-walled, relatively small, rounded cells initially forms under the periderm and eventually breaks through. Thus, the lenticel consists of a break in the tuber surface underlaid with loosely arranged, thin-walled cells. Lenticels (numbering 1–/cm^2 of surface) permit gas exchange through the relatively impervious periderm. When soils are unfavorably wet, lenticels on underground stems and tubers become enlarged (hypertrophied or proliferated) and extrude beyond the surface as white tufts approximately 0.5 mm in diameter.

Lenticels provide infection sites for several pathogens, including those inciting bacterial soft rot and late blight.

Roots

Plants from true seed produce a slender tap root that later becomes fibrous. Plants grown from seed tubers have a fibrous system of lateral roots arising usually in groups of three at the nodes of the underground stem. Lateral roots originate in the pericycle regions of roots and in meristems of the subterranean stems close to the nodal plate. Cell division in the pericycle gives rise to the root primordium, which pushes its way mechanically and possibly by enzymatic activity through the cortex. Sites of root emergence are essentially open wounds and provide infection courts for pathogens.

Selected References

ADAMS, M. J. 1975. Potato tuber lenticels: Susceptibility to infection by *Erwinia carotovora* var. *atroseptica* and *Phytophthora infestans*. Ann. Appl. Biol. 79:275–282.

ARTSCHWAGER, E. 1924. Studies on the potato tuber. J. Agric. Res. 27:809–835.

ARTSCHWAGER, E. 1927. Wound periderm formation in the potato as affected by temperature and humidity. J. Agric. Res. 35:995–1000.

BURTON, W. G. 1966. The Potato, 2nd ed. H. Veenman and Zonen, Wageningen, Holland. 382 pp.

DEAN, B. B., P. E. KOLATTUKUDY, and R. W. DAVIS. 1977. Chemical composition and ultrastructure of suberin from hollow heart tissue of potato tubers (*Solanum tuberosum*). Plant Physiol. 59:1008–1010.

DODDS, K. S. 1962. Classification of cultivated potatoes. Pages 517–539 in: D. S. Correll, ed. The Potato and Its Wild Relatives. Section Tuberarium of the Genus Solanum. Texas Res. Foundation, Renner, TX. 606 pp.

HARRIS, P. M., ed. 1978. The Potato Crop: The Scientific Basis for Improvement. Chapman and Hall, London. 730 pp.

HAYWARD, H. E. 1938. *Solanum tuberosum*. Pages 514–549 in: The Structure of Economic Plants. The Macmillan Co., New York. 674 pp.

HOWARD, H. W. 1970. Genetics of the Potato, *Solanum tuberosum*. Logos Press, Ltd., London. 126 pp.

MIZICKO, J., C. H. LIVINGSTON, and G. JOHNSON. 1974. The effects of dihydroquercetin on the cut surface of seed potatoes. Am. Potato J. 51:216–222.

OCHOA, C. M. 1962. Los *Solanum* Tuberiferos Sylvestres del Peru (Sec. *Tuberarium*, Sub-sec. *Hyperbasarthrum*). Talleres Gráficos P. L. Villanueva S.A., Lima Peru. 297 pp.

REEVE, R. M. 1974. Relevance of immature tuber periderm to high commercial peeling losses. Am. Potato J. 51:254–262.

SALAMAN, R. N. 1949. The History and Social Influence of the Potato. Cambridge University Press, London. 685 pp.

SWAMINATHAN, M. S., and H. W. HOWARD. 1953. The cytology and genetics of the potato (*Solanum tuberosum*) and related species. Bibliogr. Genet. 16:1–192.

THOMPSON, N. R. 1978. Potatoes. Pages 485–501 in: M. Milner, N.

Fig. 3. **A,** Epidermis of stem. Tuber surfaces: **B,** immature periderm; **C,** slightly more mature periderm; **D,** mature periderm of normal tuber surface; **E,** well-developed wound healing periderm on a cut surface. Bar represents 50 μm in all photographs.

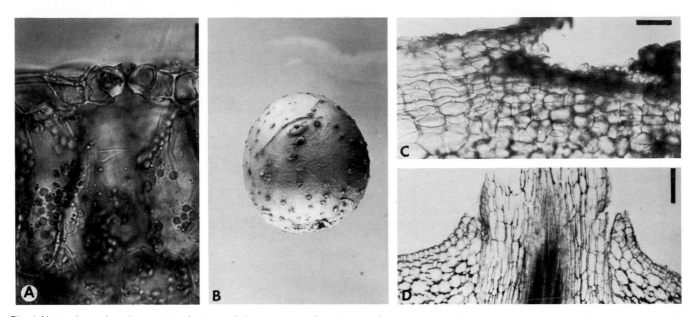

Fig. 4. Natural openings in a potato plant permitting entrance of pathogens: **A,** stomate on leaf (bar represents 20 μm); **B,** enlarged lenticels on tuber surface, usually inconspicuous but enlarged when soil is wet; **C,** section through enlarged lenticel (bar represents 100 μm); **D,** roots emerging through the surface of stems or other roots, producing open wounds (bar represents 100 μm).

S. Scrimshaw, and D. I. C. Wang, eds. Protein Resources and Technology: Status and Research Needs. Avi Publishing Co., Inc., Westport, CT.

VAN DER ZAAG, D. E. 1976. Potato production and utilization in the world. Potato Res. 19:37–72.

WIGGINTON, M. J. 1974. Effects of temperature, oxygen tension and relative humidity on the wound-healing process in the potato tuber. Potato Res. 17:200–214.

(Prepared by W. J. Hooker)

Oxygen-Temperature Relationships

Interrelationships between tuber respiration, gas exchange, and tuber temperature are operative in the field before, during, and after tuber enlargement. After harvest, they markedly influence storage life, seed performance, and market quality.

The potato tuber is capable of respiring both aerobically and, for a limited time, anaerobically. Because the natural periderm of the tuber is a barrier to gas diffusion, diffusion takes place through the lenticels. Diffusion rates differ between individual lenticels, and diffusion is further dependent upon the exposed intercellular spaces of the underlying tuber tissue. Within the tuber, gas diffusion takes place through the intercellular spaces, which occupy close to 1% of the internal tuber volume.

Oxygen (O_2) is present in the potato tuber both in the atmosphere of the intercellular spaces and dissolved in the cell sap. Carbon dioxide (CO_2) diffuses through lenticels at a rate approximately 80% of that for O_2. During early and midseason storage, CO_2 excess and O_2 deficit within the tuber are usually about equal. Later in storage CO_2 evolution may much exceed O_2 absorption.

Tuber periderm permeability is highest through immature skins during growth of the crop, decreases during maturation of the vines, and reaches a low level after death of vines. Permeability drops considerably during the first five weeks of storage and gradually rises to a level comparable to that of mature tubers in the field. A thin film of water on the tuber surface virtually stops oxygen diffusion through the lenticels and can reduce the center of the tuber to anaerobic conditions within 6 hr at 10° C and in 2 hr at 21° C.

Respiration rates of small, medium, and large tubers are essentially similar per unit of volume under satisfactory environmental conditions. However, the ratio of surface area to total volume is much higher in small tubers than in large tubers. Under conditions stimulating high respiration rates (high temperature) or reduced gas diffusion from tuber surfaces (surface water films), the smaller ratio of surface area to volume in large tubers may limit gas exchange and cause injury, whereas the larger ratio in small tubers may enable them to escape damage.

Respiration rates of immature tubers are considerably higher than those of mature tubers early in storage but later become essentially similar. Respiration of mature tubers immediately after harvest is often three times that of the same tubers a week later. This increased respiration is in part associated with mechanical injury during harvest and storage.

Respiration of potato tubers during storage (Fig. 5) is

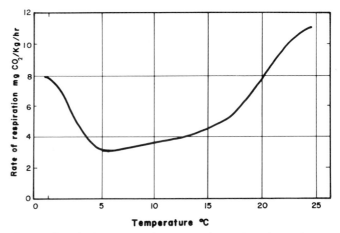

Fig. 5. Respiration rates of potato tubers at various storage temperatures. (Redrawn from W. G. Burton 1966. By permission from the author and H. Veenman en Zonen B. V.).

5

relatively high at 1 and 2° C, drops to a low at 5° C, is slightly higher at 15° C, and then rapidly rises to 25° C and above. The relatively high respiration rates within a few degrees of freezing account for problems of suboxidation such as internal mahogany browning and blackheart. The increase in respiration rates at higher temperatures contributes to internal heat necrosis and also to blackheart.

High respiration rates in freshly packaged tubers during transit deplete O_2 and release CO_2 in sufficient quantities to predispose tubers to bacterial soft rot and cause severe losses. Increasing CO_2 concentration in storage causes increased cell membrane permeability, sucrose content, and decay. Oxygen levels of 1% or lower at 14° C or higher temperatures severely impair wound healing reactions, stimulate anaerobic respiration, and increase surface mold growth, severity of decay, and blackheart. At lower temperatures these effects may not be so evident.

Selected References

BURTON, W. G. 1965. The permeability to oxygen of the periderm of the potato tuber. J. Exp. Bot. 16:16–23.

BURTON, W. G., and M. J. WIGGINTON. 1970. The effect of a film of water upon the oxygen status of a potato tuber. Potato Res. 13:180–186.

LIPTON, W. J. 1967. Some effects of low-oxygen atmospheres on potato tubers. Am. Potato J. 44:292–299.

MEINL, G. 1967. Zur Bezugsgrösse der Respirations Intensität von Kartoffelknollen. Eur. Potato J. 10:249–256.

NIELSEN, L. W. 1968. Accumulation of respiratory CO_2 around potato tubers in relation to bacterial soft rot. Am. Potato J. 45:174–181.

WIGGINTON, M. J. 1973. Diffusion of oxygen through lenticels in potato tuber. Potato Res. 16:85–87.

WORKMAN, M., E. KERSCHNER, and M. HARRISON. 1976. The effect of storage factors on membrane permeability and sugar content of potatoes and decay by *Erwinia carotovora* var. *atroseptica* and *Fusarium roseum* var. *sambucinum*. Am. Potato J. 53:191–204.

(Prepared by W. J. Hooker)

Part I.
Disease in the Absence of Infectious Pathogens

Genetic Abnormalities

Somatic mutations are of economic importance in clonally propagated potato because they may modify foliage type, tuber shape, or color of plant parts; delay maturity; and reduce crop yield. Whether yields and tuber shape are affected or not, variations are undesirable because foliage type and growth habits are principal features used to identify cultivars. A change as simple as an excess or loss of foliage vigor will cause doubt as to cultivar identity.

Wildings
These differ from normal plants by low growth, close bushy habit, numerous thin stems, reduced numbers of leaflets, large rounded terminal leaflets, almost complete absence of flowers, and increased numbers of stolons with numerous small tubers that produce many weak sprouts during storage. Yields are reduced severely by this abnormality. The small tubers may reduce the total weight of the crop by as much as 50%, but tubers of table stock size are particularly affected and, in quantity, may reduce the weight by as much as 80–90%.

Feathery Wildings
Feathery wildings bear no resemblance to true wildings except that, compared to normal plants, they produce more thin stems and many more small tubers. Plants closely resemble the normal, but the top leaflets are small, narrow, and pointed. In some cultivars, tubers have numerous eyes clustered at the apical end, producing many small, thin sprouts several weeks earlier than normal. Yield reductions from this disease are similar to those from the wilding abnormality.

Giant Hill
Plants (Plate 1) have greater height and stronger, more vigorous vines, with leaflets smaller and often coarser and thicker than those of normal plants. Tubers sprout late and plants are late in maturing, which can result in reduced yields. They have flowers and fruits in profusion and large matted roots and numerous long stolons. When allowed to mature, they produce larger and coarser tubers than do normal plants. Giant hill plants survive late blight for three weeks longer than do normal plants. Because of its late maturity, this variant is most serious in crops of early cultivars.

Although giant hill occurs in most commercial cultivars, it can be rogued with ease. However, the current practice of early foliage destruction of certified seed crops, before all giant hill plants are obvious, is probably responsible for their regular appearance in some stocks.

Tall Types
These are intermediate between giant hill and normal plants,
being a little taller, more vigorous, and two or more weeks later in maturity. Like the giant hill, the tall type is not apparent until full vine growth and may affect up to 3% in certain cultivars.

Giant hill and tall types are more frequent in the long days of higher latitudes than in lower latitude short days. Maximum yields are rarely obtained, however, because when the normal crop plants are ready to be harvested, giant hill and tall type plants are still immature. Given sufficient time for full maturity, giant hill plants may actually outyield normal plants.

Other Variations
Many other abnormal types of plants or plant parts too numerous to detail may occur in commercial crops. Examples are most russet-skinned variants (Fig. 6) and multiple-leaf, dahlia-leaf, raspberry-leaf, coarse-leaf, little-leaf, and stitched-end variants.

Nature of Variations
Most of the abnormalities described above are periclinal chimaeras caused by genetic changes, usually in the outer layers of tubers or stems. Normal plants may be recovered by excising tuber eyes and permitting growth to develop from the deeper

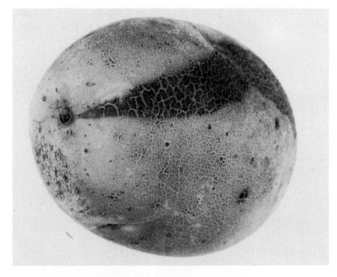

Fig. 6. Russet skin somatic mutation, in a white tuber, initiated near the stolon tip early in tuber enlargement, possibly a chimaera.

tissues. Many different kinds of variants can be produced by developing adventitious growths from callouses on tubers where eyes have been removed.

Selected References

DEARBORN, C. H. 1963. "Stitched end," "giant hill," and fasciated stem of potatoes in Alaska. Am. Potato J. 40:357–360.
HOWARD, H. W. 1967. The chimerical nature of a potato wilding. Plant Pathol. 16:89–92.
HOWARD, H. W. 1969. The chimerical nature of a feathery wilding. Eur. Potato J. 12:67–69.
HOWARD, H. W. 1970. Chimaeras. Pages 68–88 in: H. W. Howard, ed. Genetics of the Potato, *Solanum tuberosum*. Logos Press, Ltd., London. 126 pp.

(Prepared by J. Munro)

Adverse Environment

Oxygen Deficit

Oxygen requirements of underground parts of the potato during plant development are high. When oxygen concentration is reduced, stolons are abnormal and tuber development is impaired and abnormal. The degree of abnormality depends upon the severity of oxygen deficit.

Although soil compaction exerts various stresses upon underground parts of the plant, oxygen deficit may be one of the most important, resulting in delayed plant emergence, moderate to severe yield reductions, and frequently, but not always, abnormal tuber shapes. Oxygen levels within soil, root distribution, and yields are increased by cultural and tillage practices that favor improved soil porosity.

Selected References

BUSHNELL, J. 1956. Growth response from restricting the oxygen at roots of young potato plants. Am. Potato J. 33242–248.
GRIMES, D. W., and J. C. BISHOP. 1971. The influence of some soil physical properties on potato yields and grade distribution. Am. Potato J. 48:414–422.
HARKETT, P. J., and W. G. BURTON. 1975. The influence of low oxygen tension on tuberization in the potato plant. Potato Res. 18:314–319.
SOMMERFELDT, T. G., and K. W. KNUTSON. 1968. Effects of soil conditions in the field on growth of Russet Burbank potatoes in southesastern Idaho. Am. Potato J. 45:238–246.

(Prepared by W. J. Hooker)

Low Temperature Tuber Injury

Low temperature tuber injury may range from outright freezing and killing of some or all of the tuber to gradations of injury (chilling) following prolonged exposure to temperatures slightly above freezing. Tubers may be frozen in the ground before harvest or injured later by low storage temperatures. Tubers of many cultivars freeze at temperatures below $-1.7°C$. Freezing results in formation of ice crystals within the tissue, followed by rapid death. Chilling results in eventual death of cells or tissues even though the tissues may not actually have been frozen.

Symptoms

The line of demarcation between frozen and unfrozen tissue is usually distinct. Upon thawing, tissue may change progressively from dull off-white to pink and red and eventually to brown, gray or black. Frozen tissue promptly breaks down in a soft, watery rot or collapses, leaving a chalky residue as the water evaporates.

Low temperature surface injury occurs in diffuse patches as a brownish black metallic discoloration. Such tissue is predisposed to surface mold growth (Fig. 7A).

Effects of low temperature storage are primarily internal. Tuber tissue chilled to near freezing is typically a diffuse smoky gray to black and resembles certain aspects of Pythium leak.

Chilling causes formation of reducing sugars in stored tubers, resulting in a sweet flavor when cooked. Development is most rapid at temperatures slightly above freezing (0–2.5°C), progressively less severe from 2.5 to 3.5°C, and usually absent at 3.8–4.4°C. Reducing sugars cause brown discoloration in french fries or chips. Tubers stored at low temperatures frequently turn gray to black when boiled.

Chilling injury may also take the form of net necrosis, in which phloem tissue is selectively killed because it has greater sensitivity to cold than do the surrounding parenchyma storage cells (Fig. 7B). The necrotic phloem may be scattered throughout the tuber or on the chilled side or be concentrated

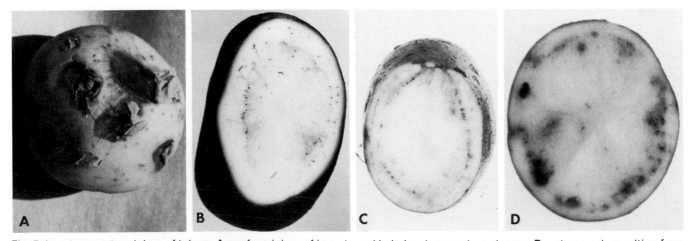

Fig. 7. Low temperature injury of tubers: **A,** surface injury of immature skin in low temperature storage; **B,** net necrosis resulting from selective killing of phloem tissue; **C** and **D,** tissue breakdown in the vascular area.

more heavily in the vascular region. Cold-induced net necrosis is very similar in appearance to virus leafroll net necrosis.

Following severe injury, blackish patches or blotches may develop near the vascular ring, which may also be partially or completely blackened (Fig. 7C and D). Injury is usually more severe near the stolon end.

Internal mahogany browning (Plate 2) is a different low temperature response, in which diffuse brownish red to black discoloration is present, usually in the central part of the tuber. This disorder grades into blackheart. Shrinkage in affected tissue results in cavities. Blackheart and probably internal mahogany browning result primarily from asphyxiation of internal tuber tissue.

Epidemiology

Individual tubers from the same lot vary considerably in response to a given temperature. Immature tubers are frequently more severely injured.

Because of hardening or acclimatization, tubers that have been stored at low temperatures are less injured by a sudden drop in temperature than are those stored at higher temperatures.

Low temperatures for a few hours or temperatures just below freezing for a short time can lower internal quality, shorten storage life, and impair suitability of the tuber for processing without leaving visible evidence. Tubers may be supercooled to approximately $-3.0°C$, and even to $-6.5°C$ for a few hours, without ice crystal formation and, if gradually warmed, do not have evident injury. However, jolting or jarring supercooled tubers will cause intracellular ice crystals to form and cells to die. Storage at low temperatures, even in the absence of symptoms, impairs the tuber's ability to form wound barriers when returned to favorable temperatures. Injured seed tubers with or without symptoms sprout poorly and may fail to produce plants because of secondary seed piece rots.

Alternating temperatures during the storage season avoids chilling injury and its associated problems. Tubers held alternately for three weeks at $0°C$ and one week at $16°C$ had a lowered content of reducing sugars (as well as of total sugars), reduced respiration rates, and no low temperature injury, whereas at constant $0°C$ tuber injuries became progressively more severe after eight weeks in storage.

Control

1) Field-frosted tubers should not be moved into storage if at all possible.

2) Hold storage temperatures at $3.5–4.5°C$, which is sufficiently high to prevent low temperature injury.

3) Maintain adequate air movement in stored potatoes to dry frost injured tubers and provide adequate oxygen for respiration.

4) Tubers injured by low temperatures or suboxidation should not be used for seed.

5) Certain cultivars are more prone to mahogany browning than are others. Varietal differences with respect to other aspects of low temperature injury are not so pronounced.

Selected References

CUNNINGHAM, H. H., M. V. ZAEHRINGER, G. BRAUSEN, and W. C. SPARKS. 1976. Internal quality of Russet Burbank potatoes following chilling. Am. Potato J. 53:177–187.

HRUSCHKA, H. W., W. L. SMITH, Jr., and J. E. BAKER. 1969. Reducing chilling injury of potatoes by intermittent warming. Am. Potato J. 46:38–53.

JONES, L. R., M. MILLER, and E. BAILEY. 1919. Frost necrosis of potato tubers. Wis. Agric. Exp. Stn. Res. Bull. 46. 46 pp.

LINK, G. K. K., and G. B. RAMSEY. 1932. Market diseases of fruits and vegetables. Potatoes. U.S. Dept. Agric. Misc. Publ. 98. 62 pp.

RICHARDSON, L. T., and W. R. PHILLIPS. 1949. Low temperature breakdown of potatoes in storage. Sci. Agric. 29:149–166.

(Prepared by W. J. Hooker)

Low Temperature Foliage Injury

Certain symptoms of nonlethal low temperature foliage injury may be confused with virus symptoms or herbicide damage. Lethal freezing of leaves and stems is readily identified.

Symptoms

Frozen leaves rapidly wilt, collapse, and when thawed, become water-soaked. They turn black when damp and brown when dried. Less severe low temperature injury, usually occurring in the early to middle part of the growing season, often produces a buff to light brown or yellow discoloration on the top of the plant and particularly at the bases of young leaflets.

Temperatures at or near $0°C$ selectively injure leaf and stem primordia and possibly cell organelles. Symptoms of this injury become evident after leaflet expansion as unilateral leaflet development, irregular distortion of leaflets, or grayish transverse banding accompanied by restricted lateral expansion (Fig. 8).

Chlorosis in diffused areas, in spots, or in portions of veins may be seen and mottle patterns may be present with or without leaf distortion following nonlethal low temperatures (Plate 3). Necrotic specks may develop on young leaves following $-0.3°C$ wet bulb temperatures. Injury of this type appears after leaves from damaged primordia have expanded. Normal growth may precede and should follow these low-temperature effects, but symptoms on injured parts persist.

Epidemiology

Low temperature injury is usually most severe in low-lying areas of fields. At high elevations and latitudes, freezing may occur at any time in the growing season.

Because leaf surfaces are frequently well hydrated and often wet with dew, wet bulb temperatures should be more reliable than dry bulb temperatures in determining critical temperatures for leaf injury.

Plants on which some leaves have been frozen recover from injury slowly, suggesting more damage than that of the tissue actually destroyed. Growth retardation may be due to resorption of tissue degradation products.

Solanum acaule, its derivatives, and approximately 10 more wild potato species, as well as several cultivated clones

Fig. 8. Leaf deformation following low temperature injury of leaf primordia.

belonging to primitive species grown in the Andes, carry considerable frost tolerance, some to as much as −5°C.

Control

1) Potato crops seldom justify frost protection such as spray irrigation, mechanical air movement, or smoke application during low temperatures.

2) Proper diagnosis of nonlethal injury is necessary in seed fields.

3) Low temperature tolerance (approaching −6°C) of hybrids involving *S. acaule* and other tuber-bearing *Solanum* species permits potato cultivation at high altitudes and possibly also at extreme latitudes.

Selected References

CHEN, P. M., M. J. BURKE, and P. H. LI. 1976. The frost hardiness of several *Solanum* species in relation to the freezing of water, melting point depression, and tissue water content. Bot. Gaz. (Chicago) 137:313−317.

ESTRADA, R. N. 1978. Breeding frost-resistant potatoes for the tropical highlands. Pages 333−341 in: H. P. Li and A. Sakai, eds. Plant Cold Hardiness and Freezing Stress. Academic Press, New York. 416 pp.

HOOKER, W. J. 1968. Sublethal chilling injury of potato leaves. Am. Potato J. 45:250−254.

LI, P. H. 1977. Frost killing temperatures of 60 tuber-bearing *Solanum* species. Am. Potato J. 54:452−456.

McKAY, R., and P. E. M. CLINCH. 1945. Frost injury simulating virus disease symptoms on potato foliage. Nature 156:449−450.

PALTA, J. P., and P. H. LI. 1979. Frost-hardiness in relation to leaf anatomy and natural distribution of several *Solanum* species. Crop. Sci. 19:656−670.

(Prepared by W. J. Hooker)

Blackheart

Blackheart results from inadequate oxygen supply for respiration (asphyxiation) of internal tuber tissue. Internal mahogany browning and internal heat necrosis grade into blackheart in severe instances and thus are, in different environments, symptoms of incipient to acute suboxidation. Blackheart was a major problem when potatoes were shipped in stove-heated railway cars.

Symptoms

Blackheart symptoms consist of black to blue-black discoloration in irregular patterns in the central portion of the tuber (Fig. 9). With acute oxygen deficiency, the whole tuber may be discolored. Demarcation at the margins is usually definite, although the black discoloration may diffuse into relatively unaffected tissue. Discolored tissue is frequently firm but on exposure to room temperatures may become soft and inky. Individual tubers vary in their responses to conditions causing blackheart.

Blackheart develops when oxygen is excluded from or unable to reach internal tuber tissue. Longer times are required for blackheart development at lower temperatures. However, blackheart develops more rapidly between 0 and 2.5°C than at 5°C. At extreme temperatures of 36−40°C or of 0°C or slightly below, blackheart may develop without oxygen exclusion because gas diffusion through the tissues is not sufficiently rapid. Tuber storage in closed bins or in deep piles without adequate aeration may result in blackheart development.

Control

1) Do not expose tubers to high temperatures nor to prolonged storage near 0°C.

2) Provide forced aeration of potatoes in closed bins.

Selected References

BENNETT, J. P., and E. T. BARTHOLOMEW. 1924. The respiration of potato tubers in relation to the occurrence of blackheart. Calif. Agric. Exp. Stn. Tech. Paper 14. 41 pp.

STEWART, F. C., and A. J. MIX. 1917. Blackheart and the aeration of potatoes in storage. N.Y. Agric. Exp. Stn., Geneva, Tech. Bull. 436:321−362.

(Prepared by W. J. Hooker)

High Temperature Field Injury

Stems may be injured at the soil line by high soil temperatures, particularly when plants are small and leaves are not large enough to shade the soil at the base of the plants. Stems typically are girdled and surfaces are tan to white, although secondary organisms may sometimes discolor the tissue to a darker brown and, in severe cases, cause rotting. Injury can also follow defoliation or vine displacement that suddenly exposes lower stems to intense sunlight. This results in a scalded apperance on the exposed side of the stem or girdling at the soil line (Fig. 10).

Tubers exposed to sunlight as they lie in the field after digging may be injured and thereby predisposed to rot in transit or storage without immediate external symptoms except possibly for watery exudates from lenticels. Intense exposure causes sunken scalded areas in a circular pattern. The threshhold of tuber flesh temperature predisposing tissue to soft rot is approximately 43°C. Such internal temperatures may exceed

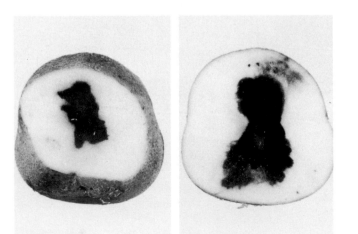

Fig. 9. Blackheart at two cross sections of the same tuber.

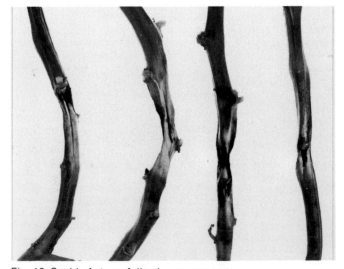

Fig. 10. Scald of stems following exposure to sun.

the air temperature when tubers are in the soil within 2.5 cm of the surface, remain on the ground after digging, or are held in bags in the sun. (See also tuber greening.)

Selected References

NIELSEN, L. W. 1954. The susceptibility of seven potato varieties to bruising and bacterial soft rot. Phytopathology 44:30–35.

(Prepared by W. J. Hooker)

Internal Heat Necrosis

Considerable confusion exists both in symptom description and in terminology for the causal factors of internal necrosis of tubers. The underlying cause of necrosis is believed to be suboxidation of rapidly respiring internal tissues during active tuber growth and high temperatures. (See also phosphorus deficiency, yellow dwarf, mop-top, and stem mottle.) In recent literature, the name Eisenfleckigkeit refers specifically to internal heat necrosis, whereas Propfenbildung and Spraing are used for stem mottle virus infections.

Symptoms
Symptoms do not develop in vines. Affected tubers usually do not show external symptoms. Necrosis may be severe toward the center of larger tubers, appearing as light tan, dark yellowish to reddish brown, or rust-colored flecks that become, in extreme cases, dark brown or even black (Fig. 11). Unusually severe symptoms may be identical to blackheart. Necrotic flecks are usually clustered off-center in the pith towards the apical end. Necrotic flecks are firm, do not break down or predispose to rot, and remain firm after cooking. Cortical tissues are seldom affected. A relationship exists between Eisenfleckigkeit and acid soils that are low in calcium. (See also calcium deficiency.)

Losses can be severe because of buyer discrimination against internal discoloration.

A somewhat similar disorder, present in Israel, produces necrotic spots in the cortex near the vascular ring and may produce interior cavities. Damage is visible from the tuber surface, with blackening of the eyes at the apical end, sunken surface spots, and a silvery sheen. No true rot develops, but many affected tubers fail to sprout. Symptoms are believed to develop in storage following high field temperature before harvest.

Histopathology
Suberin develops in walls of affected pith parenchyma cells. Cell walls first become dark at the corners. Protoplasm becomes granular and aggregates. Walls of adjacent cells also darken and finally collapse at the corners. Layers of peridermlike cells may develop outside the necrotic tissue and may isolate it. Internal pressure from periderm formation may cause the collapse of necrotic cells, but cell lysogeny has not been observed. Starch grains are generally absent in affected cells.

Epidemiology
Internal necrosis becomes progressively more severe during the growing season and is most severe during hot, dry years in light soils of sand, gravel, muck, or peat. Lack of adequate soil moisture may be as influential as high temperature in predisposing to internal necrosis. Disease is most severe in tubers near the soil surface and progressively less frequent and severe with increasing tuber depth. Straw mulch reduces soil temperature and severity of disease. In areas where the disease was formerly severe, maintenance of good vine coverage of the soil through adequate irrigation and good cultural practices has almost eliminated the problem.

Discoloration does not increase and may decrease in storage if affected tubers are not predisposed to storage rots. Transmission through affected seed tubers has not been observed, although spindly sprouts have been reported from tubers exposed to 30–40° C.

Control
1) Cultivars differ in tolerance and sensitivity.
2) Maintain vine growth adequate to shade the ground through the use of appropriate cultural practices (good fertility,

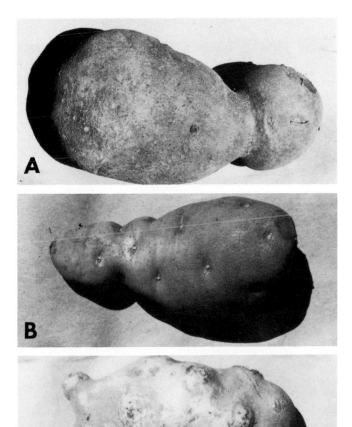

Fig. 12. Second growth: **A,** dumbbell; **B,** pointed end; **C,** protruding eyes that later form knobs. Stolon end in each case is at left.

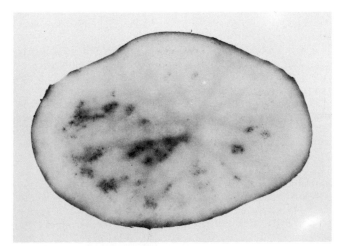

Fig. 11. Internal heat necrosis.

Fig. 13. Second growth as gemmation (**left**) and as a sprout (**right**) on an abnormally early tuber. (Left, courtesy W. M. Iritani)

adequate irrigation, and foliage protection by pesticides).

3) Do not permit tubers to remain long in the soil after vines have died.

Selected References

BRAUN, H. 1961. Die Eisenfleckigkeit der Kartoffel. Z. Pflanzenkr. Pflanzenschutz 68:542–549.

FRIEDMAN, B. A. 1955. Association of internal brown spot of potato tubers with hot, dry weather. Plant Dis. Rep. 39:37–44.

LARSON, R. H., and A. R. ALBERT. 1945. Physiological internal necrosis of potato tubers in Wisconsin. J. Agric. Res. (Washington, DC). 71:487–505.

ZIMMERMAN-GRIES, S. 1964. The occurrence of potato heat-necrosis symptoms in Israel and the use of affected tubers as seed. Eur. Potato J. 7:112–118.

(Prepared by W. J. Hooker)

Second Growth and Jelly End Rot

Second growth may be of several types: 1) deformed tubers with protruding eyes, lateral buds (knobby tubers), or apical buds (dumbbells or elongated tubers) (Fig. 12); 2) gemmation—secondary tubers on a stolon extension of the original tuber (Fig. 13 left); or 3) recently formed tubers that, before normal harvest, produce either a sprout or a leafy aboveground plant (Fig. 13 right).

Second growth is commonly attributed to high field temperatures and drought. It may, however, result from regeneration following any condition causing irregular rates of tuber development, such as uneven availability of nutrients or moisture, extremes in temperature, or vine defoliation from hail or frost. Positive separation of heat effects from drought effects is difficult because high field temperatures are usually accompanied by drought and a concomitant reduction or cessation of tuber growth. When growing conditions improve, resumption of tuber growth becomes evident as second growth.

Second growth is usually stimulated by soil temperatures of 27°C and above, although some develops at lower temperatures. Under controlled conditions, subjecting plants to 32°C for seven days was sufficient to initiate second growth. Severity was greater with longer periods of exposure and higher temperatures. Second growth was not initiated by varying the water supply alone.

Second growth and jelly end rot are interrelated because jelly end rot is prevalent in abnormally shaped tubers, particularly those with second growth. Jelly end or glassy end rot is highly seasonal in occurrence, has been reported from many potato-growing areas, and may involve 10–50% or more of the crop. Losses due to reduced tuber quality are high.

"Translucent end" or "sugar end" refers to incipient symptoms visible at harvest or developing in storage. Tubers with such symptoms frequently develop jelly end rot later. Reducing sugars in translucent end or sugar end tubers cause dark color in potato chips (Fig. 14D).

Symptoms

Stolon ends of tubers with jelly end rot become translucent to glassy, lack normal starch content, have reduced specific gravity, shrivel, and collapse into a wet jellylike substance (Fig. 14A). Jelly end rot tissue dries down to a leathery layer in dry storage (Fig. 14B). Demarcation between healthy and affected

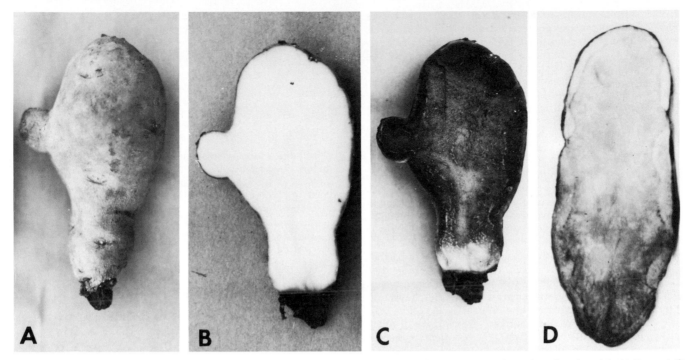

Fig. 14. Second growth: **A,** knob and jelly end rot of stolon end of tuber (bottom); **B,** longitudinal section showing dried jelly end; **C,** starch-iodine stain test demonstrating starch depletion (white area) in tissue near the jelly end portion; **D,** sugar end (stolon end, at bottom) after deep fat frying with resulting discoloration due to reducing sugars. (D, Courtesy W. M. Iritani)

tissue is distinct, with breakdown seldom extending into the tuber over 5 cm. Glassy tissue resembles that of exhausted seed pieces after plant development.

Jelly end rot is more prevalent in tubers of long-tubered cultivars, particularly those with second growth such as spindle shape (pointed ends) or dumbbell shape. Only the stolon end is affected. Round-tubered cultivars also develop jelly end rot but are not obviously deformed.

No pathogen has been consistently demonstrated in jelly end rot tissue, although populations of secondary organisms, particularly bacteria, are usually high.

The stolon end of the tuber normally has the highest starch content. In second growth, starch deposited in early tuber development is apparently hydrolyzed and translocated from the stolon end to the apical end (Fig. 14C).

When second growth takes the form of gemmation, tubers (particularly round varieties) develop in chains on stolons. Carbohydrates are translocated from the primary tuber into the secondary tuber, which is of normal quality. Internal quality changes in the primary tuber (glassy texture and low specific gravity) become apparent after foliage of the plant has died. Because they show no external symptoms, such tubers cannot be readily separated during grading and cause considerable loss in quality and market value of the crop.

Epidemiology

Incidence of jelly end rot is strongly influenced by seasonal conditions, being a severe problem in some years and not a problem in others. It is related to high field temperatures accompanied by drought or insufficient irrigation in the early growing period, followed by favorable temperature or sufficient water to stimulate tuber growth. When tubers that are immature at harvest because of early drought stress are stored immediately after harvest at low temperatures (5.5°C), they are predisposed to increased incidence of translucent ends, as compared to those stored at higher temperatures (9°C).

Control

1) Irrigate adequately to maintain uniform growing conditions throughout the season, particularly during early tuber development.

2) Avoid cultivars with second growth characteristics.

Selected References

FRIEDMAN, B. A., and D. FOLSOM. 1953. Potato tuber glassy end and jelly end rot in the Northeast in 1948 and 1952. Plant Dis. Rep. 37:455–459.

IRITANI, W. M., and L. WELLER. 1973. The development of translucent end tubers. Am. Potato J. 50:223–233.

LUGT, C., K. B. A. BODLEANDER, and G. GOODIJK. 1964. Observations on the induction of second-growth in potato tubers. Eur. Potato J. 7:219–227.

MURPHY, P. A. 1936. Some effects of drought on potato tubers. Emp. J. Exp. Agric. 4:230–246.

NIELSEN, L. W., and W. C. SPARKS. 1953. Bottleneck tubers and jelly-end rot in the Russet Burbank potato. Idaho Agric. Exp. Stn. Res. Bull. 23. 24 pp.

(Prepared by W. J. Hooker)

Hollow Heart

Hollow heart is associated with excessively rapid tuber enlargement. It presents a serious problem because, lacking external symptoms, the defect usually becomes apparent only when the tuber is cut in half. Incidence of hollow heart is highest in the largest tubers and in certain lots may affect up to 40% of the tubers by weight.

Symptoms

Usually one cavity forms near the center of the tuber. In many cultivars, cavities are lens or star shape and angular at the corners. They appear as a splitting within the tuber (Fig. 15), and internal walls of the cavities are either white or light tan to straw color. In other cultivars, but less commonly, cavities are round to irregular in shape.

Before hollow heart develops, central tissue may be water-soaked or translucent; a brown necrotic patch appears early in tuber formation in some cultivars. Rot seldom starts at hollow heart sites, although in rare instances mold may be present.

Histopathology

Three types of origin are described: 1) a necrotic patch, up to 1 cm in diameter, composed of many single cells or small groups of cells, becomes enclosed by periderm, then turns brown, shrinks, and collapses to produce a cavity; 2) necrotic starch-free cells differentiate, causing a brown spot approximately 1 mm in diameter, often in the center of very small tubers, and produce a cavity that enlarges with tuber growth and is surrounded by a partially suberized cambium layer; 3) internal tissue tensions cause splitting, which results in a lens-shaped cavity not preceded by cell necrosis. Translocation of substances from the central portion of the tuber and resorption in other parts of the plant may also be involved.

Epidemiology

Hollow heart is most severe during growing seasons or under cultural practices favoring rapid tuber enlargement. Rapidly growing tubers have a relatively higher incidence of hollow heart than do those that grow more slowly. Moisture stress (deficiency) followed by conditions favoring rapid growth predispose the tuber to hollow heart.

Hollow heart is frequently severe in fields with poor stands where plants are irregularly spaced. Practices that inhibit rapid tuber growth or that stimulate large numbers of small tubers, such as close spacing of plants, reduce incidence of hollow heart. Marginal potassium deficiency may be a factor in hollow heart predisposition, particularly in cultivars prone to the disorder. Increasing potassium fertility over that required for normal growth reduces hollow heart incidence.

Identification for Marketing

X-ray examination of whole tubers under water effectively identifies the condition without tuber destruction. Removal of large tubers and those with low specific gravity is only partially effective in eliminating hollow heart potatoes before marketing.

Control

1) Potato cultivars differ in severity of incidence and in the type of internal cavity produced.

2) Close and regular spacing of plants increases competition and prevents excessively rapid tuber enlargement, which usually reduces incidence of hollow heart.

3) Avoid missing hills in planting, and use sound cultural practices to assure good stands.

4) Maintain uniform soil moisutre levels to stimulate uniform tuber growth rates.

5) Additional potassium fertilization reduces incidence of hollow heart even though total yields may not be increased.

Selected References

CRUMBLY, I. J., D. C. NELSON, and M. E. DUYSEN. 1973. Relationships of hollow heart in Irish potatoes to carbohydrate reabsorption and growth rate of tubers. Am. Potato J. 50:266–274.
FINNEY, E. R., Jr., and K. H. NORRIS. 1978. X-ray scans for detecting hollow heart in potatoes. Am. Potato J. 55:95–105.
NELSON, D. C. 1970. Effect of planting date, spacing, and potassium

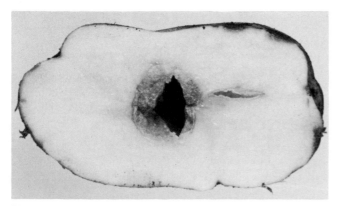

Fig. 15. Hollow heart, showing transverse and longitudinal splitting.

on hollow heart in Norgold Russet potatoes. Am. Potato J. 47:130–135.
VON WENZL, H. 1965. Die histologische Unterscheidung dreier Typen von Hohlherzigkeit bei Kartoffelknollen. Z. Pflanzenkr. Pflanzenschutz 72:411–417.

(Prepared by W. J. Hooker)

Surface Abrasions

Immature tubers that are mechanically injured during harvest before the periderm is mature exhibit feathering, i.e., shreds of loose skin exposing underlying flesh (Fig. 16). The wound may heal under optimum conditions but frequently dehydrates, becomes somewhat sunken, and turns dark brown with a sticky surface due to bacterial growth. Such tubers do not store well.

Mature tubers may be skinned by rough handling during harvesting and grading operations, thus providing infection courts for wound pathogens. Sunken scald spot develops when fresh wounds are dehydrated, especially after tubers are allowed to stand for some time in direct sunlight or desiccating wind before storage. Surface discoloration of wounds, with associated rot problems, also follows low temperature storage before wound healing is complete. Such tubers may become flaccid from dehydration.

Control

1) Avoid mechanical damage at every stage of the digging, harvesting, and grading operations.

2) Protect tubers from sunlight and heat; avoid excessive dehydration before storage.

3) Provide optimum storage conditions until wounds are completely healed.

Selected References

SMITH, W. L. Jr. 1952. Effect of storage temperatures, injury, and exposure on weight loss and surface discoloration of new potatoes. Am. Potato J. 29:55–61.
WHITEMAN, T. M., and J. M. LUTZ. 1954. Sunken scald spot field injury evident in stored potatoes. Am. Potato J. 31:43–49.

(Prepared by W. J. Hooker)

Tuber Cracks

Tuber cracking is of four types: 1) growth cracks from internal pressure, 2) growth cracks from virus infection, 3) mechanically produced cracks, and 4) harvest cracks.

Growth cracking (bursting) usually follows the long axis of the tuber and results from internal pressure exceeding the tensile strength of surface tissues during tuber enlargement. High internal turgor pressure develops from tissue expansion during rapid tuber growth. Fertilizer placed so that growth is excessively rapid increases growth cracking. Growth cracks in

Fig. 16. Immature tuber surfaces, skinned and abraded.

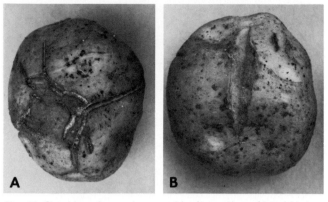

Fig. 17. Cracking: **A,** crack caused by impaction of turgid tuber during digging; **B,** growth crack that healed over before digging.

Fig. 18. Harvest or thumbnail cracks following mild bruising and surface drying. (Courtesy W. C. Sparks)

the field before harvest frequently wound heal and, as tubers continue to grow, become relatively shallow and of little consequence (Fig. 17B). Wound-healed cracks seldom become infected. Potato cultivars differ in susceptibility to injury.

Growth cracks may also develop in tubers of plants with the yellow dwarf virus, potato mop-top virus, or certain strains of the spindle tuber viroid.

Mechanical cracking during harvesting may follow sudden impacts (Fig. 17A). Cracking is dependent upon varietal response, tuber maturity, internal tuber turgor, and degree of mechanical compression during harvest and movement into storage. Immature tubers and large tubers are most easily injured. Severity is greatest when tuber temperatures are low and tissue is turgid. Tubers with high internal turgor are easily cracked to a depth of 5 mm or more. Extreme turgidity results when soil moisture levels are high and roots continue to function after vines have been suddenly killed by frost, by herbicides, or by harvesting tubers when vines are green. Root pruning, undercutting, or pre-harvest vine killing reduces incidence of cracking. Delaying digging for a few hours early in the day until the soil is warmed may also reduce cracking considerably. Severely cracked tubers are of little value because wound healing is incomplete, dehydration is rapid, and incidence of tuber rot may be high. Paradoxically, shatter bruising becomes more severe with high tuber turgor and black spot intensifies with low tuber turgor.

Harvest cracks are crescent shaped, resembling cracks made with a thumbnail. They are usually shallow, 1–2 mm deep, and result from rough handling and drying of the tuber surface tissue after digging, particularly when tubers are turgid (Fig. 18). Severity of injury depends on intensity of bruising and rapidity of dehydration. Direct harvesting by machinery often reduces incidence of harvest cracks as compared to harvesting by lifting tubers to the soil surface and gathering them later.

The catechol test (1–1.5% practical grade) reveals areas of injury by turning them dark red to purplish within 3–5 min of treatment. Catechol, a polyphenol, is oxidized by phenolase, which is released from recently broken cells. It is not effective in identifying tissue bruised by black spot without rupture of the surface, nor is the reaction obtained after wounded tissue has healed.

Control

1) Little can be done to avoid cracking during the growing period except by judicious irrigation, fertilizer application, plant spacing, and cultivar selection.

2) Delay harvest until vines have been dead for some time and tuber periderm has matured. Avoid harvesting from cold soil.

3) Avoid sudden impact on tubers, and protect them from rapid drying after digging and during transit from field to storage.

Selected References

IRITANI, W. M. 1968. The use of catechol for enhancing bruise detection. Am. Potato J. 45:312.

PAINTER, C. G., and J. AUGUSTIN. 1976. The effect of soil moisture and nitrogen on yield and quality of the Russet Burbank potato. Am. Potato J. 53:275–284.

SMITTLE, D. A., R. E. THORNTON, C. L. PETERSON, and B. B. DEAN. 1974. Harvesting potatoes with minimum damage. Am. Potato J. 51:152–164.

WERNER, H. O., and J. O. DUTT. 1941. Reduction of cracking of late crop potatoes at harvest time by root cutting or vine killing. Am. Potato J. 18:189–208.

(Prepared by W. J. Hooker)

Blackspot

Blackspot is always caused by bruising injury, either from impact during harvest, handling, and grading, or from pressure during storage. The disorder is well known in North America and northern Europe and has become an increasingly serious problem in most potato-growing areas that have adopted mechanical harvesting and handling techniques.

Symptoms

Blue-gray to black discolored areas develop just beneath the tuber skin (Fig. 19). Internal symptoms do not appear immediately after brusing but develop to full intensity over a period of 1–3 days as flattened, spheroidal blue-gray patches centering in the vascular region. Margins are diffuse and grade into the unaffected tissue. Blackspot is usually more noticeable at the stolon end of the tuber than at the apical end. Tubers with internal blackspot frequently show no external symptoms.

Melanin is present on intracellular protoplast surfaces and on inner wall surfaces of affected cells. The absence of wound periderm in lesions is characteristic of blackspot and helps to differentiate this disorder from other internal defects such as internal brown spot, heat necrosis, and certain internal lesions caused by pathogens.

Histopathology

Bruising injury initiates a series of biochemical oxidations in damaged cells. Phenyl substrates such as tyrosine are oxidized to conjugated quinones by polyphenol oxidases. The quinones polymerize to produce the black pigment. Oxidation reactions are usually completed within 24 hr of bruising, and spots neither enlarge nor disappear during subsequent storage.

Epidemiology

Severity of blackspot is determined by both the number of damaged cells and the amount of melanin produced in each one. An impact will damage more cells in susceptible tubers than it will in resistant ones, and larger, deeper spots will form. Tuber susceptibility is influenced by a number of factors.

Tubers with low turgor pressure are more likely to have severe blackspot. Therefore, blackspot is usually more serious after a dry growing season in nonirrigated growing areas. Conditions such as low soil moisture, poor root development, or hot dry

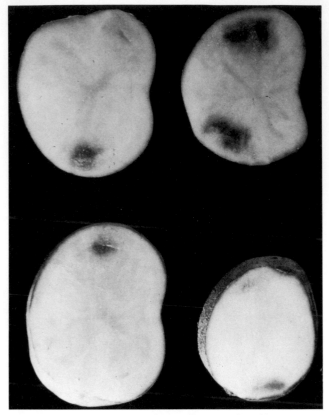

Fig. 19. Blackspot internal bruising. (Courtesy S. L. Sinden and R. W. Goth)

days before harvest tend to predispose to bruising injury. Because of tuber hydration differences, tubers with high specific gravity are usually more susceptible to bruising than are tubers with low specific gravity from the same lot. Susceptibility can increase during storage because of physiological aging and dehydration.

Mature tubers are more susceptible than immature tubers, and the stolon end is more susceptible than the apical end. Temperature of the tubers at the time of bruising influences severity. Tubers bruised at 20–30° C are less affected by blackspot than are those bruised at temperatures below 10° C. Because of differences in both mechanical strength and solids content, cultivars differ significantly in susceptibility to bruising and blackspot development.

Tubers harvested from soils deficient in potassium tend to be more susceptible to bruising and blackspot development. Low potassium content in tubers is associated with high phenolic content and low tuber hydration. High phenolic content and active oxidase systems in damaged cells result in abundant production of melanin.

Nitrogen fertilization, ethylene concentrations, and soil carbon dioxide levels have been reported to affect blackspot susceptibility in some growing areas. The specific effect of any one environmental factor on susceptibility of tubers to blackspot depends on the cultivar, the cultural conditions, and the interaction with other environmental factors.

Control

1) Reduction of bruising is most important for control of blackspot. Use equipment for harvesting, transporting, grading, and handling tubers that is well designed and carefully adjusted to minimize impact forces.

2) Use sound cultural management practices, including adequate potash fertilization, especially on heavy soils that are likely to be deficient in potassium. Irrigate as long as vines are green.

3) Warm tubers in storage to 20° C before grading and other handling operations. Using sprout inhibitors and adding moisture to the storage atmosphere will help prevent dehydration and bruise damage in tubers taken out of storage.

4) Use resistant cultivars.

Selected References

DWELLE, R. B., G. F. STALLKNECHT, R. E. McDOLE, and J. J. PAVEK. Effects of soil potash treatment and storage temperature on blackspot bruise development in tubers of four *Solanum tuberosum* cultivars. Am. Potato J. 54:137–146.
KUNKEL, R., M. L. WEAVER, and N. M. HOLSTAD. 1970. Blackspot of Russet Burbank potatoes and the carbon dioxide content of soil and tubers. Am. Potato J. 47:105–117.
SCHIPPERS, P. A. 1971. Measurement of black spot susceptibility of potatoes. Am. Potato J. 48:71–81.
SMITH O. 1968. Internal black spot of potatoes. Pages 303–307 in: O. Smith, ed. Potatoes: Production, Storing, Processing. Avi Publishing Co., Inc., Westport, CT. 642 pp.
TIMM, H., M. YAMAGUCHI, D. L. HUGHES, and M. L. WEAVER. 1976. Influence of ethylene on black spot of potato tubers. Am. Potato J. 53:49–56.

(Prepared by S. L. Sinden and R. W. Goth)

Tuber Greening and Sunscald

When tubers are exposed for some time to light in the field or after harvest, chlorophyll forms in the leucoplasts and tuber tissue turns green. Sun green, sometimes less correctly called sunscald, develops in tubers not covered by soil in the field and therefore exposed to intense sunlight.

Green tissue may extend 2 cm or more into the tuber and is often accompanied by purple pigmentation. Such tissue is high in solanine, bitter in flavor, and believed to be toxic to humans when ingested. The processes of greening and solanine production are independent. Affected tubers are not marketable, and losses may be high.

Sunscald injury develops in tubers exposed to intense sunlight as restricted areas with almost-white skin, often covering a sunken necrotic area. (See high temperature field injury.)

Certain potato cultivars have a tendency to set tubers near the soil surface. Throwing soil toward the plants during cultivations often effectively covers tubers and reduces greening. However, tubers may be exposed later by soil erosion or by cracks formed as soil dries or tubers enlarge. Ordinarily, severely greened tubers are not predisposed to rot unless sunlight and heat have been intense.

Table stock potatoes should be stored in the dark. Fluorescent or natural lighting in market displays causes superficial and, occasionally, deeper layers of the tuber to turn green. Color is persistent; it is not removed by placing tubers in the dark. Greening develops more rapidly at room temperature than in cold storage. Potato cultivars show differences in intensity of greening and the depth to which it develops. Tuber rinses with surfactants, used experimentally, show promise of reducing the intensity of greening.

Selected References

AKELEY, R. V., G. V. C. HOUGHLAND, and A. E. SHARK. 1962. Genetic differences in potato-tuber greening. Am. Potato J. 39:409–417.
GULL, D. D., and F. M. ISENBERG. 1960. Chlorophyll and solanine content and distribution in four varieties of potato tuber. Proc. Am. Soc. Hortic. Sci. 75:545–556.
POAPST, P. A., I. PRICE, and F. R. FORSYTH. 1978. Controlling post storage greening in table stock potatoes with ethoxylated mono- and diglyceride surfactants and an adjuvant. Am. Potato J. 55:35–42.

(Prepared by W. J. Hooker)

Internal Sprouting

Sprouts that develop during storage may become ingrown by penetrating into the tuber. Internal sprouts frequently are in eyes with tightly clustered multiple "rosette" sprouts, which may be unbranched or, more frequently, branched (Fig. 20). Sprouts may penetrate the tuber directly above, or sprouts from an eye on the bottom of the tuber may grow up through the same tuber. Sprouts from tubers with deep eyes may penetrate into the side of the eye depression.

The disorder has been known for over a century. It is more frequent in old tubers and in those stored at 12–15°C. Pressure on tubers within the storage pile restricts sprout growth and induces sprout penetration of tuber tissue. In old tubers, sprouts often tuberize within the parent tuber, splitting it open.

Internal sprouting was recently associated with sprout inhibitors used in concentrations below those required for complete sprout inhibition. Concentrations that completely inhibit all external sprouts also inhibit internal sprouts, but insufficient concentrations actually stimulate internal sprouts. Isopropyl-*m*-chlorocarbanilate (CIPC) stimulates internal sprouting to a greater extent than does pressure on tubers under a deep pile. Other chemicals stimulating tightly clustered, multiple sprouts also cause internal sprouting.

Necrosis at or slightly below the sprout apex is common on the external sprouts of tubers containing internal sprouts, and apicies of internal sprouts become similarly necrotic when they emerge from the tuber. (See Ca deficiency.)

Selected References

EWING, E. E., J. W. LAYER, J. C. BOHN, and D. J. LISK. 1968. Effects of chemical sprout inhibitors and storage conditions on internal sprouting in potatoes. Am. Potato J. 45:56–71.

WIEN, H. C., and O. SMITH. 1969. Influence of sprout tip necrosis and rosette sprout formation on internal sprouting of potatoes. Am. Potato J. 46:29–37.

(Prepared by W. J. Hooker)

Secondary Tubers

Tubers sprout either in storage or in the field, producing new tubers directly without forming a normal plant. Secondary tubers form on sprouts from physiologically old tubers after completion of the rest period when carbohydrate reserves are low (Fig. 21). The disorder is associated with warm (20°C) storage followed by low temperature after planting or by transfer of sprouted tubers from warm to cold storage. Even at low temperatures, however, physiologically overmature tubers held past normal usage form secondary tubers. Usually the problem is of minor importance, although poor field stands with missing hills result. (See also calcium deficiency.)

Selected References

BURTON, W. G. 1972. The response of the potato plant and tuber to temperature. Pages 217–223 in: A. R. Rees, K. E. Cockshull, D. W. Hand, and R. G. Hurd, eds. Crop Processes in Controlled Environments. Academic Press, New York. 391. pp.

DAVIDSON, T. M. 1958. Dormancy in the potato tuber and the effects of storage conditions on initial sprouting and on subsequent sprout growth. Am. Potato J. 35:451–465.

VAN SCHREVEN, D. A. 1956. On the physiology of tuber formation in potatoes. I. Premature tuber formation. Plant Soil 8:49–55.

(Prepared by W. J. Hooker)

Coiled Sprout

The disease has been reported primarily from the British Isles, where up to 26% of plants in certain fields are affected, but it probably exists elsewhere.

Symptoms

Underground sprouts lose their normal negative geotropic habit and coil, sometimes rather tightly, with the curved portion of the stem often swollen and sometimes fasciated or split (Fig. 22). Light brown lesions with transverse or longitudinal cracks may be present on the stem inside the coil. Delay in emergence of coiled sprouts results in uneven stands. Affected plants may produce more stems than normal, and tubers may form unusually early and mature slowly.

Causal Factors

Coiling is believed to be the result of overmature seed, soils resistant to sprout penetration and emergence, or infection by a fungus. *Verticillium nubilum* Pethybridge has been isolated from affected stems. This pathogen has caused superficial browning and russeting of some stem bases, accompanied by shallow cortical invasion underlaid by suberin. In some instances, inoculation with the pathogen has caused coiled sprout, but *V. nubilum* is not the sole cause of the disease.

Low soil temperatures, presprouting in light, long sprouts at

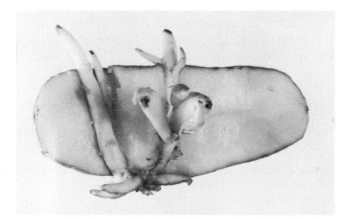

Fig. 20. Internal sprouting, showing rosette of sprouts on the underside, small tubers on internal sprouts, and necrosis under or at the sprout apices. The last resembles calcium deficiency. (Courtesy E. E. Ewing)

Fig. 21. Secondary tubers formed directly on sprouts from physiologically old tubers.

Fig. 22. Coiled sprout. (Courtesy M. A. Alli, J. H. Lennard, and A. E. W. Boyd)

the time of planting, overmature seed tubers with long sprouts that may form tubers before sprouts emerge from the soil, and deep planting in compacted soil have been associated with the disorder. Ethylene in low concentrations as produced by sprouts induces coiled sprout characteristics.

Control

1) Avoid planting seed tubers with long sprouts.
2) Avoid planting in compacted soil resistant to sprout penetration.

Selected References

ALI, M. A., J. H. LENNARD, and A. E. W. BOYD. 1970. Potato coiled sprout in relation to seed tuber storage treatment and to infection by *Verticillium nubilum* Pethybr. Ann. Appl. Biol. 66:407–415.
CATCHPOLE, A. H., and J. R. HILLMAN. 1975. Studies of the coiled sprout disorder of the potato. Parts 2 and 3. Potato Res. 18:539–545 and 597–607.
COX, A. E. 1970. Early tuberisation associated with coiled sprout in Craigs Royal potatoes. Plant Pathol. 19:49–50.
TIMM, H., J. C. BISHOP, J. W. PERDUE, D. W. GRIMES, R. E. VOSS, and D. N. WRIGHT. 1971. Soil crusting on potato plant emergence and growth. Calif. Agric. 25(August):5–7.

(Prepared by W. J. Hooker)

Hair Sprout

Tubers with hair (or spindle) sprout germinate early, sometimes even before harvest, producing thin sprouts as small as 2 mm in diameter. A single tuber may produce normal sprouts and hair sprouts from different eyes. Hot, dry conditions in the late growing season, particularly during tuber development, favor hair sprout formation. Hair sprout has been induced in tubers of certain, but not all, cultivars by warm water treatment for 2 hr at 45°C.

Virus infection has not been consistently associated with spindle sprouts. Early maturity following attack by *Colletotrichum atramentarium* may predispose to spindle sprout. Certain mycoplasma diseases (aster yellows) and pysllid yellows are also known to cause hair sprout. (See also genetic abnormalities—wildings and feathery wildings—and internal heat necrosis.)

Selected References

ORAD, A. G., and F. P. SAN ROMAN. 1955. Conditions which determine spindling sprout of potato in Spain. Pages 160–170 in: E. Streutgers, A. B. R. Beemster, D. Noordam, and J. P. H. Van der Want, eds. Proc. Second Conf. Potato Virus Dis., Lisse-Wageningen, June 1954, H. Veenman and Zonen, Wageningen. 193 pp.
SNYDER, W. C., H. E. THOMAS, and S. J. FAIRCHILD. 1946. Spindling or hair sprout of potato. Phytopathology 36:897–904.
STEINECK, O. 1955. Untersuchungen und Beobachtungen über die Fadenkeimigkeit von Kartoffelknollen. Phytopathol. Z. 24:195–210.
WENZL, H. 1966. Fadenkeimigkeit und Kallose-bildung durch Warmwasserbehandlung von Kartoffelknollen. Rev. Roum. biol. Ser. Bot. 1:271–276.

(Prepared by W. J. Hooker)

Nonvirus Leafroll

Leafrolling is a symptom with several unrelated causes. When carbohydrate translocation from the foliage is impaired, starch accumulates in the leaves, causing them to become leathery and roll upward (Fig. 23) in a way similar to that of virus leafroll. Leafroll-like symptoms, with or without chlorosis and red pigmentation, may accompany Rhizoctonia, Fusarium wilts, and other diseases; injury by mycoplasmas; and mechanical injury of stems.

Leafrolling may also be genetic. Although genetic nonvirus leafroll has symptoms similar to those of virus leafroll, its nonvirus origin was established by failure of graft transfers to infect suitable indicator hosts. The recessive mutant gene (lr) causes starch accumulation in leaves, but anatomically defective phloem is not detectable.

Certain nutritional soil conditions, such as nitrogen toxicity, also cause nonvirus leafroll. Rolling of leaves is uniformly intense from plant to plant. In contrast, leafroll severity in virus leafroll usually differs considerably between plants. Cultivars differ in degree of response. Correction of unsatisfactory soil

Fig. 23. Nonvirus leafroll: **A**, plant grown in calcareous muck; **B**, apparently normal plant grown in similar soil supplemented with sulfur. (Courtesy W. J. Hooker and G. C. Kent)

conditions permits normal leaf development.

Toproll affects the plant's apical leaves. Its symptoms are similar to those of virus leafroll, but toproll results from feeding by the potato aphid *Macrosiphum euphorbiae* in the absence of the leafroll virus. When aphid feeding is discontinued, new growth is normal. Plants grown from progeny tubers are free from toproll and give normal yield.

Selected References

DZIEWONSKA, M. A., and R. K. McKEE. 1965. Apparent virus infection in healthy potatoes. Potato Res. 8:52.

GIBSON, R. W. 1975. Potato seed tubers do not transmit top-roll. Plant Pathol. 24:107–108.

HOOKER, W. J., and G. C. KENT. 1950. Sulfur and certain soil amendments for potato scab control in the peat soils of northern Iowa. Am. Potato J. 27:343–365.

LeCLERG, E. L. 1944. Non-virus leafroll of Irish potatoes. Am. Potato J. 21:5–13.

REESTMAN, A. J. 1973. Some observations on "toprol" in the Netherlands. Proc. Trienn. Conf., Eur. Assoc. Potato Res. 5:184.

SIMMONDS, N. W. 1965. Mutant expression in diploid potatoes. Ann. Appl. Biol. 20:65–72.

VOLK, G. M., and N. GAMMON, Jr. 1954. Potato production in Florida as influenced by soil acidity and nitrogen sources. Am. Potato J. 31:83–92.

(Prepared by W. J. Hooker)

Hail Injury

Hail tears and often perforates leaves (Plate 4). Although the potato plant has a remarkable ability to recover from foliage damage, hail may cause defoliation sufficiently severe to impair yields. On stems, injury is localized at the point of impact; epidermal tissue turns gray with a paperlike sheen.

Yield reduction varies with severity of injury, time of injury, and cultivar. Greatest losses result from vine damage within 2–3 weeks after blossom set. Marketable yields are adversely affected through the relative increase in small tubers or in off-shaped tubers. Specific gravity may be reduced when hail damages mature vines. Hail injury seldom predisposes foliage to infection except for the *Ulocladium* disease in the tropics.

Selected Reference

BERESFORD, B. C. 1967. Effect of simulated hail damage on yield and quality of potatoes. Am. Potato J. 44:347–354.

(Prepared by W. J. Hooker)

Wind Injury

Wind injury is evident on upper surfaces of leaves that, through wind movement, have been rubbed by other leaves, usually those directly above the affected area. Discolored tissue is brown when dry, varies in size, has a glistening or oily appearance (Plate 5), and sometimes extends through the leaf. following severe high winds, leaves may be tattered at the edges and the plant may appear hard (non succulent). Very cold winds lasting for some time cause undersurfaces of leaves, particularly those turned over by the wind, to be brown, sometimes with a silvery or glassy sheen.

Leaf injury may be more severe at the edges of the field.

During harvest, tubers in sacks in the field may be damaged by drying wind. Injury may become evident later during storage as sunken spots underlying the skinned portions of tubers. The surfaces may be overgrown with bacterial slime, causing decay in storage. Damage is greater with immature than with mature tubers and with open mesh sacks than with tightly woven ones.

Selected References

GRACE, J. 1977. Plant Response to wind. Academic Press, New York. 204 pp.

WHITEMAN, T. M., and J. M. LUTZ. 1954. Sunken scald spot field injury evident in stored potatoes. Am. Potato J. 31:43–49.

(Prepared by W. J. Hooker)

Lightning Injury

Lightning injury frequently accompanies severe thunderstorms. Severity is influenced by field and plant hydration. Diagnosis of injured plants without information about the distribution of plant injury in the field is hazardous because plant symptoms may resemble those of blackleg or *Rhizoctonia* injury, and tuber symptoms may closely resemble those of ring rot. Bleached stems at soil line often resemble those injured by high temperatures.

Symptoms

Within a few minutes to a few hours following lightning, stems collapse and tops of plants irreversibly wilt. Leaves may remain green and turgid even though stems collapse. In most cases, injury extends 5–10 cm or more above the soil line; it rarely proceeds from the top downward. Affected portions are soft, water-soaked, and discolored black to brown. Tissues soon dry and become chocolate brown to tan, with the surface layer light tan to almost white. Pith collapse results in flattened, ribbed, or angular stems with longitudinal depressions of the surface. Collapsed pith forms horizontal plates, and when the stem is split longitudinally the pith appears crosshatched or ladderlike (Fig. 24A).

Leaf petioles in contact with soil are often collapsed.

Belowground portions of stems and roots frequently escape injury. Water transport is often sufficient to maintain green, turgid tops, which may survive despite collapse of parenchymatous stem tissue.

Injured tubers have brown to black skin necrosis with some cracking. Surfaces may be injured on opposite sides of the tuber, with intermediate cortex and pith becoming soft and remaining relatively normal in color for a short time. Later, decay extends completely through the tuber, leaving a hole. Unaffected portions of tubers often remain solid. In some, the cortex is intact even though the pith is completely collapsed and watery. Tubers may resemble those affected by other diseases such as Pythium leak or bacterial ring rot (Fig. 24B).

Epidemiology

A field may show various patterns of injury: 1) areas in which all plants are killed adjacent to clearly defined areas of healthy plants, 2) areas with dead plants at the center and progressively reduced injury toward the periphery, 3) poor defined areas with no focus of injury, in which plants with varying degrees of injury are scattered among dead plants and unaffected ones (Fig. 25), and 4) a number of scattered, relatively small foci containing several plants in various stages of injury.

These variations result from differences in intensity of electrical discharge and from variations in soil hydration. Electrostatic discharge and conduction along irrigation pipes is related to field distribution of injured plants.

Selected References

HOOKER, W. J. 1973. Unusual aspects of lightning injury in potato. Am. Potato J. 50:258–269.

SAMUEL, G. 1940. Lightning injury to potato tubers. Ann. Appl. Biol. 27:196–198.

WEBER, G. F. 1931. Lightning injury of potatoes. Phytopathology 21:213–218.

(Prepared by W. J. Hooker)

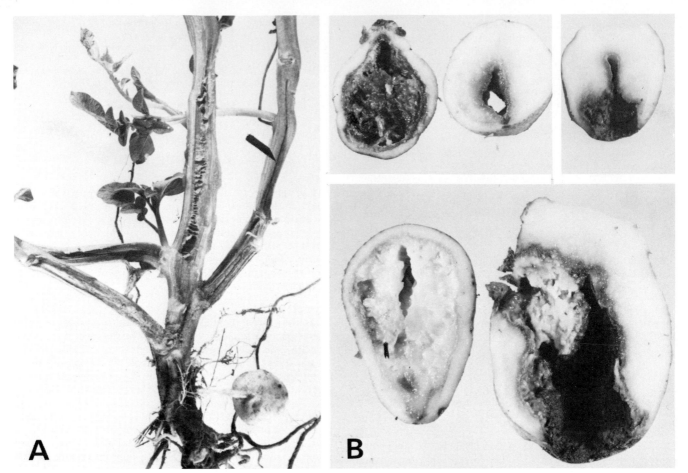

Fig. 24. Lightning injury: **A,** typical injury on plant, with necrosis at stem base, creaselike collapse of a branch (arrow), and ladderlike breakdown of pith; **B,** tuber injury, sometimes resembling ring rot. Note hole through one tuber.

Fig. 25. Lightning damage, showing irregularly affected area.

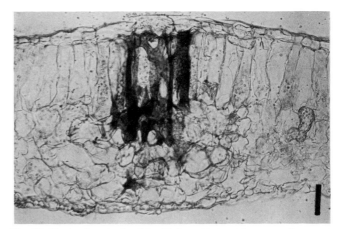

Fig. 26. Photochemical oxidant injury, first evident as collapse of palisade mesophyll. Bar represents 50 μm.

Air Pollution: Photochemical Oxidants

Potato injury by photochemical oxidants such as ozone, probably peroxyacetyl nitrate, and related compounds has recently been recognized. Symptom differences in the field between the several photochemical oxidants have not yet been determined. Losses in sensitive cultivars may be severe following exposure early in the season.

Symptoms

Upper leaf surfaces are stippled by darkly pigmented spots, sometimes with chlorosis and often with a bronzed appearance (Plate 6). Injury is most severe on the lower, older leaves and progresses upwards. Later, plants become generally chlorotic, with premature leaf death progressing usually from the bottom upward. Leaves eventually drop but do not abscise rapidly.

Lower leaf surfaces may be light in color occasionally with a glazed or silvery sheen.

Except for severe reduction in yield associated with very early senescence, tuber symptoms have not been reported.

Symptoms become evident within 24 hr following heavy exposure, but symptoms of advanced necrosis and chlorsis may require 10–14 days.

Histopathology

Palisade mesophyll cells are first affected (Fig. 26), becoming water-soaked and later necrotic. The spongy mesophyll and epidermis collapse later.

Epidemiology

Oxidant injury is present in North America along the Atlantic coast, in the Great Lakes region, the southeastern states, and the Pacific southwest.

Photochemical oxidants accumulate under two conditions: when relatively large areas of high atmospheric pressure are present or when air masses stagnate under a layer of warm air over cool land surfaces. Episodes in which normal dispersion of air pollutants is prevented may occur infrequently during the growing season. Extent of injury is influenced by the concentration of oxidants, length and frequency of exposure, plant genotype, and stage of plant growth.

Field exposures of approximately 0.15 ppm ozone for a day or two are usually sufficient to injure exposed foliage. The amount of damage is largely influenced by density of the foliage mass. If the foliage mass is sufficiently large, a "sink" effect is produced, by which air pollutants are absorbed or adsorbed by leaf surfaces and removed from the immediate environment, thereby protecting nearby foliage. Thus, exposed leaves above the foliage canopy may be severely damaged and leaves within the canopy escape injury. Injury may be more severe at field margins than in the center. If plants are small, leaf and stem exposure is complete and the sink effect is negligible; thus plant injury may be severe following an early season episode.

Ozone injury can predispose potato leaves to *Botrytis cinerea* infection and may increase susceptibility to other pathogens.

Control

Wide varietal differences in tolerance exist.

Cultural practices stimulating vigorous early season vine growth may hasten plants past the susceptible, small vine stage. Maintenance of a heavy foliage canopy until the tuber crop is assured may lessen or avoid midseason injury.

Selected References

BRASHER, E. P., D. J. FIELDHOUSE, and M. SASSER. 1973. Ozone injury in potato variety trials. Plant Dis. Rep. 57:542–544.

HEGGESTAD, H. E. 1973. Photochemical air pollution injury to potatoes in the Atlantic Coastal States. Am. Potato J. 50: 315–328.

HOOKER, W. J., T. C. YANG, and H. S. POTTER. 1973. Air pollution injury of potato in Michigan. Am. Potato J. 50:151–161.

MANNING, W. J., W. A. FEDER, I. PERKINS, and M. GLICKMAN. 1969. Ozone injury and infection of potato leaves by *Botrytis cinerea.* Plant Dis. Rep. 53:691–693.

MOSLEY, A. R., R. C. ROWE, and T. C. WEIDENSAUL. 1978. Relationship of foliar ozone injury to maturity classification and yield of potatoes. Am. Potato J. 55:147–153.

(Prepared by W. J. Hooker)

Air Pollution: Sulfur Oxides

Although potato leaves are relatively resistant to injury by sulfur oxides, they respond with interveinal necrotic areas that are light tan to white (Plate 7), and yields may be reduced. Injury should be anticipated in areas with air flow drainage patterns downwind from power plants and smelters. If sulfur oxides are injuring potatoes, symptoms on nearby sensitive plants (alfalfa, bean, soybean, beet, *Amaranthus* spp., bindweed, morning glory, lettuce, curly dock, plantain, ragweed, or sunflower) should confirm the diagnosis.

Selected References

JONES, H. C., D. WEBER, and D. BALSILLIE. 1974. Acceptable limits for air pollution dosages and vegetation effects: Sulfur dioxide. Paper No. 74-225. Air Pollution Control Assoc. 67th Annual Meeting, Denver.

THOMAS, M. D., and R. H. HENDRICKS. 1956. Effect of air pollution on plants. Section 9, pages 1–44 in: P. L. Magill, F. R. Holden, and C. Ackley, eds. Air Pollution Handbook. McGraw-Hill, New York.

(Prepared by W. J. Hooker)

Chemical Injury

A wide range of chemicals accidentally or improperly applied can cause divergent symptoms on foliage and in tubers, with severity depending upon the nature of the chemical, its dosage, environmental factors, and plant maturity and variety. Vine-killing preharvest defoliants frequently cause necrosis at the stolon attachment and vascular discoloration of the stem end, resembling symptoms of stem-end browning or Verticillium wilt. Interveinal leaf tissues may be burned. Moisture stress increases symptom severity.

Growth-regulating herbicides for weed control in potatoes or herbicides airborne from nearby areas may cause leaf distortion superficially suggesting virus infection (Fig. 27A and B). Tuber skin color may be affected. Some (2,4,5-trichlorophenoxyacetates) cause necrosis not unlike that from severe deep scab and also tuber deformation (Figs. 27C and D).

In storage, netting of tuber surfaces and dehydration have followed foliage application of maleic hydrazide; abnormal sprouting has been associated with other compounds. (See internal sprouting.)

Improper application of fertilizer to foliage or application too close to the seed piece in the soil causes foliage or seed tuber necrosis, followed by decay, poor stands, and low plant vigor.

Selected References

FRYER, J. D., and R. J. MAKEPEACE, eds. 1972. Weed Control Handbook. Vol. 2. Recommendations Including Plant Growth Regulations, 7th ed. Blackwell Scientific Publications, London. 424 pp.

HOOKER, W. J., and A. F. SHERF. 1951. Scab susceptibility and injury of potato tubers by 2,4,5-trichlorophenoxyacetates. Am. Potato J. 28:675–681.

MUNSTER, J., and P. CORNU. 1971. Dégáts internes causés aux tubercules de pommes de terre par la sécheresse ou par l'application de reglone. Rev. Suisse Agric. 3:55–59.

MURPHY, H. J. 1968. Potato vine killing. Am. Potato J. 45:472–478.

POABST, P. A., and C. GENIER. 1970. A storage disorder in

Fig. 27. Chemical injury: **A,** burn of interveinal leaf tissue; **B,** Leaf deformation by growth-regulating herbicide; **C** and **D,** tuber injury from foliage application of 2,4,5-trichlorophenoxyacetate. (C and D, Courtesy W. J. Hooker and A. F. Sherf)

Kennebec potatoes caused by high concentrations of maleic hydrazide. Can. J. Plant Sci. 50:591–593.

STEPHENS, H. J. 1965. The place of herbicides in the potato crop. Eur. Potato J. 8:33–51.

(Prepared by W. J. Hooker)

Stem-End Browning

Stem-end browning describes an internal, brown discoloration of tuber tissue near the stem end or stolon attachment. In its broadest sense, the term is applied to a shallow discoloration with one or more unknown causes. These may include chemical injury, sudden death of the vines, or infection from pathogens, including viruses. Usually it does not show up at harvest but develops during the first one to three months in storage. It is often confined to the 12-mm section at the stolon end of the tuber and is more frequent in smaller tubers. If penetration is deeper, the brown strands are confined to the xylem of the vascular ring.

The disorder may be confused with virus leafroll net necrosis (Fig. 28A). However, the latter penetrates deeper into the tuber, and its brown necrotic strands involve the phloem either inside or outside of the xylem. It is more prevalent in larger tubers, and affected tubers always produce leafroll-infected plants.

Recently, stem-end browning has been associated with early season leafroll virus infection, in which a limited amount of tissue is affected and the virus does not always establish infection. Thus affected tubers may not produce leafroll plants.

Stem-end browning may also come from another virus, as yet unidentified. The virus is apparently graft-transmitted and is more prevalent in certain clone lines; its characteristics have not been further clarified.

Stem-end browning has no apparent effect on yield, although affected tissue is believed to be sterile. Stem-end browning is distinguished from Fusarium or Verticillium wilts by culture techniques and from Verticillium infection or defoliants by darker color and coarser strands. Other tuber discolorations often confused with stem-end browning include those due to rapid vine killing (by chemicals or flame) or to frost or frost necrosis.

Control

Cultivars differ in resistance.

Avoid excessive applications of phosphorus, potassium, or chlorine in the fertilizer.

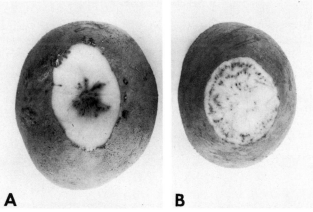

Fig. 28. **A,** Stem-end browning; **B,** virus leafroll net necrosis. (Courtesy Main Life Sciences and Agricultural Experiment Station)

Selected References

FOLSON, D., and A. E. RICH. 1940. Potato tuber net-necrosis and stem-end browning studies in Maine. Phytopathology 30:313–322.

MANZER, F. E., D. C. MERRIAM, and R. H. STORCH. 1977. Effects of time of inoculation with PLRV on internal tuber necrosis. Am. Potato J. 54:476 (Abstr.).

RICH, A. E. 1950. The effect of various defoliants on potato vines and tubers in Washington. Am. Potato J. 27:87–92.

RICH, A. E. 1951. Phloem necrosis of Irish potatoes in Washington. Wash. Agric. Exp. Stn. Bull. 528.

ROSS, A. F. 1946. Susceptibility of Green Mountain and Irish Cobbler commercial strains to stem-end browning. Am. Potato J. 23:219–234.

ROSS, A. F. 1946. Studies on the cause of stem-end browning in Green Mountain potatoes. Phytopathology 36:925–936.

ROSS, A. F., J. A. CHUCKA, and A. HAWKINS. 1947. The effect of fertilizer practice including the use of minor elements on stem-end browning, net necrosis, and spread of leafroll virus in the Green Mountain variety of potato. Maine Agric. Exp. Stn. Bull. 447:96–142.

(Prepared by A. E. Rich)

Nutrient Imbalance

Nutrient deficiencies or excesses are frequently difficult to diagnose without analysis of the plant and may be confused with other environmental stresses. Deficiency or excess of a particular nutrient may be influenced by its availability, its balance with other nutrients, soil pH, ion-exchange levels, and other factors. Because potatoes are grown under widely different conditions of altitude, day length, light intensities, soil types, temperatures, and soluble salts, general symptoms described for nutrient disorders are often based upon controlled sand culture trials rather than on field conditions.

Potatoes grow well in soils above pH 5.0. In more acid conditions, Ca and Mg deficiency, phosphate fixation, and ammonium, Mn, and Al toxicity may occur, and leaching of some nutrients (e.g., Mg) is increased. Conversely, highly calcareous or overlimed soils create unfavorable alkaline reactions that reduce availability of Mn, Fe, B, and Zn.

General References

CHAPMAN, H. D., ed. 1966. Diagnostic Criteria for Plants and Soils. University of California, Division of Agricultural Science. 793 pp.

HOUGHLAND, G. V. C. 1964. Nutrient deficiencies in the potato. In: Sprague, H. B., ed. 1964. Hunger Signs in Crops, 3rd ed. David McKay Co., Inc., New York. 461 pp.

WALLACE, T. 1961. The Diagnosis of Mineral Deficiencies in Plants by Visual Symptoms, a Colour Atlas and Guide, 3rd ed. Chemical Publishing Co., Inc., New York. 125 pp. with 312 color plates.

Nitrogen

Adequate N in the presence of sufficient P and K stimulates apical and lateral meristems and thus increases leaf development. Adequate N should be available during rapid plant growth and tuberization. N requirements increase rapidly with plant growth, as N is translocated from lower to upper leaves and much of it eventually to the tubers.

Deficient plants are generally chlorotic, slow growing, erect, and have small, erect, pale green leaves (Plate 8). Lower leaves are most severely affected. Veins stay green somewhat longer than does interveinal tissue. The extent of deficiency determines severity of stunting, chlorosis, loss of lower leaves, and yield reduction.

Speckle leaf, brown to black spots about 1 mm in diameter that may coalesce on lower leaves of some early cultivars, is particularly severe following heavy rainfall or irrigation and is alleviated by nitrogen side dress.

When N toxicity occurs, yields are reduced; root development is poor; and leaves may roll upward or be deformed as "mouse ear." N toxicity can result from the form of N available to the plant. Ammonium and/or nitrites formed from urea and diammonium phosphates are toxic. In certain soil conditions, principally very acid soils, conversion of ammonium nitrogen to nitrate nitrogen is impaired. In nutritional leafroll, nitrate nitrogen is insufficient to balance an otherwise normal amount of ammonium nitrogen available to the plant.

Surface applications of urea, especially when banded at high rates, can cause damage from ammonia volatilization. Burning of leaves and stem lesions developing near urea pellets are due to ammonia volatilization and not to an osmotic or salt effect.

Selected References

MEISINGER, J. J., D. R. BOULDIN, and E. D. JONES. 1978. Potato yield reductions associated with certain fertilizer mixtures. Am. Potato J. 55:227–234.

VITOSH, M. L., and R. W. CHASE. 1973. Speckle leaf of potato as affected by fertilizer and water management. Am. Potato J. 50:311–314.

Phosphorus

P is essential early in plant growth and later in tuberization. Early season deficiency retards growth of terminals, and plants are small, spindly, and somewhat rigid. Leaflets fail to expand normally, are crinkled or cup-shaped (Plate 8), darker than normal, lusterless, and may be scorched at the margins. Lower leaves may drop. Leaflets are not bronzed. Leaf petioles are more erect than normal. Maturity may be delayed.

Roots and stolons are reduced in both number and length. Tubers lack external symptoms, but internal rusty brown necrotic flecks or spots are scattered throughout the flesh in sometimes radial patterns. (See also internal heat necrosis and calcium deficiency.)

Deficiency occurs on a wide range of soil types: calcareous soils, peat or muck, light soils with low initial P content, and heavy soils in which P is fixed. Much of the P is translocated from vines to tubers, and the crop removes a considerable amount of P from the soil. Banding of P lateral to the seed piece decreases P fixation and improves P uptake over that from broadcast application. Little can be done to alleviate P-deficiency symptoms during the growing season, although foliage applications with neutral ammonium phosphate or polyphosphate are helpful.

Where P levels are very high, especially in alkaline soils, the uptake and/or utilization of Zn or Fe may be reduced.

Selected References

BAERUG, R., and K. STEENBERG. 1971. Influence of placement method and water supply on the uptake of phosphorus by early potatoes. Potato Res. 14:282–291.

HOUGHLAND, G. V. C. 1960. The influence of phosphorus on the growth and physiology of the potato plant. Am. Potato J. 37:127–138.

Potassium

K is essential for normal growth and is highly mobile within the plant.

Early appearance of unusually dark green, bluish green, or glossy foliage is a dependable symptom of K deficiency. Light

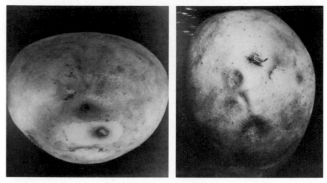

Fig. 29. Potassium deficiency symptoms on tubers. Note corky sunken areas at stolon end. (Courtesy W. M. Laughlin)

green spots (approximately 1 mm in diameter) appear between veins of larger leaflets, resembling mild mosiac. When K is in relatively short supply, older leaves first become bronzed, then necrotic (Plate 9), and senesce early. Leaflet margins from the middle to the top of the plant roll upward. Leaflets are small, cupped, crowded together, crinkled, and bronzed on the upper surface. The overall bronzed effect of the foliage is predominant. Leaves frequently have dark brown specks on the lower surface, which may coalesce and cause marginal necrosis. Symptoms may develop rapidly within four days during sunny, bright weather following cloudy, rainy periods. Necrosis is severe and may superficially resemble early blight. Stalks may be slender with short internodes. When K is acutely deficient, the growing point is affected and general dieback develops. Plants become short and squatty with shortened internodes. They appear droopy because of downward leaf curling.

Roots are poorly developed and stolons are short. Tuber size and yield are reduced. Necrotic, brown, sunken lesions develop at stolon ends of tubers of plants with necrotic foliage. Later, the affected tissue dries out, forming a hollow spot, 2 mm or more in diameter, surrounded by corky tissue (Fig. 29).

K deficiency predisposes to black spot. During early storage, K-deficient tubers frequently develop brown to black enzymatic discoloration of raw cut surfaces on exposure to air. Discoloration is frequently more severe at the stolon end of the tuber. Tuber flesh also becomes dark after cooking.

K deficiency is most common on light, easily leached, sand, muck, or peat soils. Exchangeable K should exceed 200 kg/ha (178 lb/A) in the upper 20 cm of soil.

Selected References

BAERUG, R., and R. ENGE. 1974. Influence of potassium supply and storage conditions on the discoloration of raw and cooked potato tubers of cv. Pimpernell. Potato Res. 17:271–282.

FONG, K. H., and A. ULRICH. 1969. Growing potato plants by the water culture technique. Am. Potato J. 46:269–272.

LAUGHLIN, W. M. 1966. Effect of soil applications of potassium, magnesium sulfate and magnesium sulfate spray on potato yield, composition and nutrient uptake. Am. Potato J. 43:403–411.

LAUGHLIN, W. M., and C. H. DEARBORN. 1960. Correction of leaf necrosis of potatoes with foliar and soil applications of potassium. Am. Potato J. 37:1–12.

MULDER, E. G. 1949. Mineral nutrition in relation to the biochemistry and physiology of potatoes. Plant and Soil 2:59–121.

Calcium

Ca-deficient plants are spindly, with small, upward rolling, crinkled leaflets having chlorotic margins that later become necrotic (Plate 9). In severe deficiency, leaves are wrinkled and stem tips cease to function, giving a rosette appearance. Root meristems cease to grow.

Tubers on Ca-deficient plants develop diffuse brown necrosis in the vascular ring near stolon attachments, and later similar

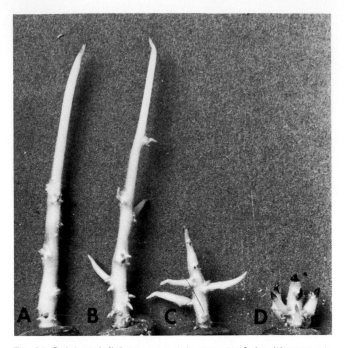

Fig. 30. Calcium deficiency on potato sprouts: **A,** healthy sprout; **B,** sprout treated with calcium sulfate; **C** and **D,** calcium-deficient sprouts. Note necrosis of tips and tendency for lateral branching. (Courtesy P. W. Dyson and J. Digby, and copyright permission of Macaulay Institute for Soil Research)

flecks form in the pith. Tubers may be extremely small. Internal rust spot is more severe on dry soils with low Ca, a tendency towards acidity, and a moderate to low base-exchange capacity. (See also internal heat necrosis and phosphorus deficiency.)

Seed tubers in Ca-deficient soil remain hard and produce relatively normal roots. Sprouts become necrotic immediately behind the tip and fail to grow (Fig. 30). In storage, sprouts become necrotic 3–5 mm below the tip due to collapse of outer cortex and inner pith and, later, of vascular tissue. Multiple lateral branches form below the sprout tips, and, with certain cultivars, small tubers known as "little potato" form prematurely before the development of aboveground sprouts. Ca deficiency and internal sprouting show certain relationships.

Symptoms are most severe on sandy soils below pH 5.0, where symptoms of Mn or Al toxicity may also be present. Calcium treatment of sprouts reduces incidence of necrosis below the tip. Liming the soil above pH 5.2 should be avoided because of potential problems with common scab.

Transfer of Ca from old leaves to young leaves and from the top of the plant to the tubers is limited. Ca must therefore be available during the entire growing period, particularly during tuberization.

Selected References

BRAUN, H., and D. E. WILCKE. 1967. Bodenprofile und ihre Beziehungen zum standortbedingten Auftreten der Eisenfleckigkeit bei Kartoffeln. Phytopathol. Z. 59:305–336.
DeKOCK, P. C., P. W. DYSON, A. HALL, and F. B. GRABOWSKA. 1975. Metabolic changes associated with calcium deficiency in potato sprouts. Potato Res. 18:573–581.
WALLACE, T., and E. J. HEWITT. 1948. Effects of calcium deficiency on potato sets in acid soils. Nature 161:28.

Magnesium

Mg deficiency is one of the most commonly encountered nutritional problems. Because Mg is highly mobile within the plant, new growth appears essentially normal and symptoms develop on older leaves. A pale, light green color—later, a more definite necrosis—begins at the leaf tips and margins and

progresses between the veins, becoming most severe toward the center of the leaf. Leaves are usually thick and brittle and roll upward, with tissue raised between the veins (Plate 10). Necrotic leaves either hang on the plant or abscise. Roots are stunted, reducing the ability of the plant to absorb Mg.

Mg deficiency usually occurs on sandy acid soils that are readily leached, but may occur on heavier soils. High rates of K fertilizer or high K levels in soil accentuate Mg deficiency. Solubility of Mg is increased by acid-forming fertilizer. Symptoms frequently follow leaching after periods of heavy rain. Exchangeable Mg should exceed 50 ppm for mineral soils and higher (100 ppm) for muck soil.

Mg may be supplied as $MgSO_4$ in fertilizer or dolomitic limestone or as a 2% $MgSO_4$ foliage spray. Higher concentrations usually may be applied to foliage without injury.

Selected References

BONDE, R. 1934. Potato spraying—The value of late applications and magnesium-bordeaux. Am. Potato J. 11:152–156.
CHUCKA, J. A., and B. E. BROWN. 1938. Magnesium studies with the potato. Am. Potato J. 38:301–312.
SAWYER, R. L., and S. L. DALLYN. 1966. Magnesium fertilization of potatoes on Long Island. Am. Potato J. 43:249–252.

Sulfur

S deficiency, as reported in several locations in Wisconsin on Planefield loamy sand, is a general yellowing with a slight upward rolling of leaflets. Symptoms vary from slight to marked chlorosis over the entire plant. Beneficial responses have been obtained with sulfur soil applications or with fertilizers containing sulfur.

Aluminum

Al toxicity causes roots to become short and stubby with few branches. Leaves remain normal in color although plants are small and spindly, with branches rising at acute angles. Potato is relatively tolerant to Al toxicity.

Al solubility is often high in soils below pH 5.0. Soil conditions may be corrected by adding superphosphate fertilizers, increasing soil pH to 5.5 or above with lime, or increasing the organic content of the soil.

Selected References

BROWN, B. A., A. HAWKINS, E. J. RUBINS, A. V. KING, and R. I. MUNSELL. 1950. Causes of very poor growth of crops on a formerly productive soil. Soil Sci. Soc. Am. Proc. 15:240–243.
HAWKINS, A., B. A. BROWN, and E. J. RUBINS. 1951. Extreme case of soil toxicity to potatoes on a formerly productive soil. Am. Potato J. 28:563–577.

Boron

In plants with B deficiency, growing points die; lateral buds become active; internodes are shortened; leaves thicken and roll upward; and the plant assumes a bushy appearance. Starch accumulation in leaves is pronounced and may resemble virus leafroll. Roots are short, thick, and stunted. Tubers are small, showing surface cracking particularly at the stolon end and localized brown areas under the skin near the stolon end or brown vascular discoloration.

Some sandy soils, peat soils, and overlimed, acid, upland, podzolized soils are inherently low in B or apparently fix B so that it becomes unavailable to plants.

Applications of B should be made cautiously because B is toxic to potatoes in relatively small amounts and deficiency is rare.

Selected Reference

MIDGLEY, A. R., and D. E. DUNKLEE. 1940. The cause and nature of overliming injury. VT Agric. Exp. Stn. Bull. 460. 22 pp.

Zinc

Zn deficiency causes stunting of plants and upward rolling of young, chlorotic leaves suggestive of early virus leafroll symptoms, with terminal leaves being somewhat vertical. Gray brown to bronze areas, later becoming necrotic, may develop on leaves near the middle of the plant and later involve all leaves. Brownish spots may develop on petioles and stems.

Zn deficiency in recently developed land, alkaline soils, or soils with excessively high P results in severe stunting, leaf malformation, and indistinct bronzing or yellowing around leaf margins. The "fern leaf" symptom (Fig. 31) is present when the youngest leaves are cupped upwards and rolled, becoming thick, brittle, and puckered from expansion of intercostal tissue and apparent lack of expansion of leaf margins. Severely affected plants die early.

Foliage applications of $ZnCl_2$ or $ZnSO_4$ alleviate deficiency symptoms. In some cases without recognizable Zn deficiency, yield increases have been obtained when Zn salts were applied to foliage as fungicides or used as soil treatments. Excessive liming or application of P enhances symptoms of Zn deficiency.

Speckle bottom (small to large necrotic spotting and chlorosis of basal leaves, which progress upward) has responded to applications of zinc.

In laboratory studies, Zn toxicity develops as general stunting, with a slight chlorosis at tips and margins of upper leaves and purple coloration on the undersides of lower leaves.

Fig. 31. "Fern leaf" symptom of zinc deficiency. (Courtesy L. C. Bowan and G. E. Leggett)

Selected References

BOAWN, L. C., and G. E. LEGGETT. 1963. Zinc deficiency of the Russet Burbank potato. Soil Sci. 95:137–141.

CIPAR, M. S., D. E. HUNTER, W. W. WEBER, R. MILLER, and P. PORTER. 1974. Soil fumigation and zinc status of soils in relationship to potato speckle-bottom disease development and control. Potato Res. 17:307–319.

HOYMAN, W. G. 1948. Potato-fungicide experiments in 1948. N.D. Agric. Exp. Stn. Bimonthly Bull. 11:32–35.

LANGILLE, A. R., and R. I. BATTEESE, Jr. 1974. Influence of zinc concentration in nutrient solution on growth and elemental content

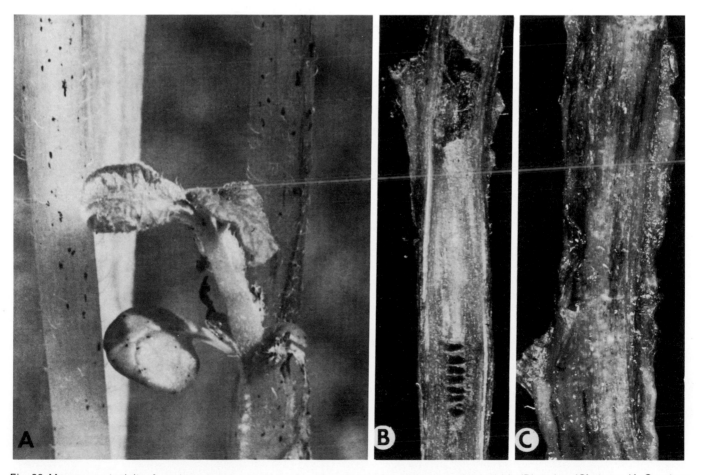

Fig. 32. Manganese toxicity: **A,** early symptoms on stems; later symptoms of manganese toxicity within **(B)** and on **(C)** stems. (A, Courtesy H. W. Gausman; B and C, courtesy K. C. Berger)

of the "Katahdin" potato plant. Am. Potato J. 51:345–354.

SOLTANPOUR, P. N., J. O. REUSS, J. G. WALKER, R. D. HEIL, L. W. LINDSAY, J. C. HANSEN, and A. J. RELYEA. 1970. Zinc experiments on potatoes in the San Luis Valley of Colorado. Am. Potato J. 47:435–443.

Manganese

Deficiency symptoms develop on the upper parts of the plant as loss in luster. Light green interveinal chlorotic tissue later becomes yellow to white (Plate 10). Lower leaves are least affected, but leaves near shoot tips often roll upward. When the deficiency is severe, brown necrotic spots develop along the veins of younger leaves.

Mn deficiency is possibly the most common micronutrient problem for potatoes grown on muck, sandy muck, or depressional soils in central and eastern coastal areas of the United States. It is reported on calcareous or excessively limed soils of high pH. Mn should be applied when leaf tissue tests show less than 25 ppm Mn. Manganese sulfate applied to foliage at the rate of 1.1–2.2 kg of Mn per hectare (1–2 lb/A) is useful to correct the deficiency. Certain fungicides containing Mn also alleviate the condition.

Mn toxicity, to which the potato is especially sensitive, has been called stem streak, stem streak necrosis, land streak, or stem break. Cultivars differ in sensitivity or tolerance. Early season Mn toxicity develops slowly; initial symptoms are necrotic flecking of stem and petioles (Fig. 32A). Sometimes leaf flecks develop into elongate, dark brown pitted streaks. This occurs first at the lower stem and progresses upward, being most severe at petiole bases and developing on the petioles (Fig. 32B and C). Necrosis becomes evident at 400 ppm in lower leaves. It appears first on the epidermis and later extends deep into the cortex, ray tissue, and pith. The Mn content in leaf tissue and the symptom severity increase rapidly after the blossom stage. Affected parts become necrotic and dark brown and are extremely brittle. The terminal bud may eventually die. The plant remains stunted and may die early. In solution culture, 25 ppm Mn reduces growth.

Leaves lose their typical bright green color and show a pale, yellow-green interveinal chlorosis that becomes progressively severe, often with marginal necrosis. Eventually the leaves dry, hang down, and break off as the petiole becomes brittle. Inverveinal necrosis may precede leaf death.

Symptoms have not been described in tubers except that yield may be severely impaired.

Neither pH alone, Ca deficiency, Mg deficiency, nor Al toxicity cause stem streak necrosis. Stem necrosis by rugose mosaic virus may be superficially somewhat similar.

Symptoms of Verticillium wilt are enhanced in pot culture in soil with a high level of Mn.

Mn increases in solubility as the soil becomes more acid; toxicity occurs on light acid soils at pH 5.0 and below. Additions of lime to raise soil above pH 5.0 are usually effective in avoiding injury. Both symptom severity and Mn content of leaves are reduced with Ca lime soil treatment and increased with chloride or sulfate fertilizers.

Selected References

BERGER, K. C., and G. C. GERLOFF. 1947. Stem streak necrosis of potatoes in relation to soil acidity. Am. Potato J. 24:156–162.

CHENG, B. T., and G. J. OUELLETTE. 1968. Effect of various anions on manganese toxicity in *Solanum tuberosum*. Can. J. Soil Sci. 48:109–115.

LANGILLE, A. R., and R. I. BATTEESE, Jr. 1974. Influence of manganese concentration in nutrient solution on the growth and elemental content of the "Katahdin" potato plant. Can. J. Plant Sci. 54:375–381.

ROBINSON, D. B., G. D. EASTON, and R. H. LARSON. 1960. Some common stem streaks of potato. Am. Potato J. 37:67–72.

WHITE, R. P., E. C. DOLL, and J. R. MELTON. 1970. Growth and manganese uptake by potatoes as related to liming and acidity of fertilizer bands. Soil Sci. Soc. Am. Proc. 34:268–271.

WHITE, R. P., A. R. SIETING, and E. C. DOLL. 1972. Manganese fertilization of potatoes in Presque Isle County. Mich. State Univ. Agric. Exp. Stn. Res. Rep. 179. 2 pp.

(Prepared by W. J. Hooker with assistance from L. M. Walsh on nitrogen and sulfur sections and from R. E. Lucas on magnesium and manganese sections)

Part II.
Disease in the Presence of Infectious Pathogens

Bacteria

Blackleg, Bacterial Soft Rot

Blackleg affects stems and may produce soft rot in tubers. Blackleg and bacterial soft rot are principally caused by two varieties of the same species of bacterium, *Erwinia carotovora*. They are found wherever potatoes are grown. Bacterial soft rot also affects the fleshy and leafy organs of a wide range of other plants and *E. carotovora* var. *atroseptica* has been reported on sunflowers in Mexico and sugar beets in the United States.

Symptoms

Blackleg. Symptoms occur at any stage of plant development. Stems of infected plants typically exhibit an inky black decay, which usually begins at the decaying seed piece and may extend up the stem only a few millimeters or for its entire length (Fig. 33A, Plate 11). Stem pith is often decayed above the black discoloration, and vascular tissues in the stem are often discolored. Infected plants are commonly stunted and have a stiff, erect growth habit, particularly early in the season. Foliage becomes chlorotic, and leaflets tend to roll upward at the margins. Leaflets and, later, entire plants may wilt, slowly decline, and eventually die. Young shoots may be invaded and killed before emergence.

Stems, petioles, and leaves may also become infected through wounds such as petiole scars, hail, or wind damage. Infection may progress up or down the stems or petioles, thus producing typical blackleg symptoms on plants that do not show infection from infected seed pieces. In wet weather, decay is soft and slimy and may spread to most of the plant. Under dry conditions, infected tissues become dry and shriveled and are often restricted to the underground portion of the stem.

Tubers produced by infected plants may show symptoms ranging from slight vascular discoloration at the stolon end to soft rot of the entire tuber. Typically, infected tubers have soft rot in the pith or medullary region of the tuber only, extending into the tuber from the stolon end (Fig. 33B and C).

Soft Rot. Tubers can also be affected with soft rot while in storage or in the soil before harvest, and seed tubers decay after planting. Infection occurs through lenticels and wounds or through the stolon end of the tuber via the infected mother plant. Lesions associated with lenticels appear as slightly sunken, tan to brown, circular water-soaked areas, approxi-

Fig. 33. Blackleg: **A,** *Erwinia carotovora* var. *atroseptica* infection progressing up stems from decayed seed piece; **B** and **C,** tuber infection through stolon from plant infected with blackleg.

mately 0.3–0.6 cm in diameter (Plate 12). In dry environments they may become sunken, hard, and dry. Sometimes infection is arrested, and the diseased area dries, leaving a sunken area filled with a mass of hard, black, dead material. Lesions associated with injuries are irregular in shape, sunken, and usually dark brown.

Soft rotted tissues are wet, cream to tan, with a soft, slightly granular consistency (Plate 13). Infected tissues are sharply delineated from healthy ones and are easily washed away. Brown to black pigments often develop near the margins of lesions. Rotting tissue is usually odorless in the early stages of decay but develops a foul odor and a slimy or ropy consistency as secondary organisms invade infected tissue.

Causal Organisms

Blackleg. *E. carotovora* var. *atroseptica* (Van Hall) Dye and sometimes *E. carotovora* var. *carotovora* (Jones) Dye cause blackleg. *E. chrysanthemi* Burkholder, McFadden, & Dimock has recently been isolated from infected potato plants with blackleg symptoms in Peru.

Soft Rot. *E. carotovora* var. *carotovora* (Jones) Dye and *E. carotovora* var. *atroseptica* (Van Hall) Dye are the most common causes of soft rot. Some pectolytic *Pseudomonas* spp., *Bacillus* spp., *Clostridium* spp., and *Flavobacterium pectinovorum* have also been found associated with soft rot infections.

E. carotovora is easily cultured and produces deep pits, or craters, on selective media that contain polypectate, such as Stewarts MacConkey-pectate medium or the Cuppels and Kelman crystal violet-pectate medium (CVP). Some *Pseudomonas* spp. and *F. pectinovorum* can also be cultured on these media but produce only very shallow pits. *Bacillus* spp. cannot be cultured on the CVP medium.

E. carotovora var. *atroseptica* and *E. carotovora* var. *carotovora* are rod-shaped, Gram-negative bacteria, approximately 0.7×1.5 μm in size, and have peritrichous flagella. They are nonspore-forming and are facultatively anaerobic.

E. carotovora var. *atroseptica* forms acid from maltose and α-methylglucoside, produces reducing substances from sucrose, and does not grow above 36°C on nutrient agar or in nutrient broth.

Typical strains of *E. carotovora* var. *carotovora* do not form acid from α-methylglucoside nor reducing substances from sucrose and will not grow above 36°C on nutrient agar or in nutrient broth. Most strains do not produce acid from maltose, although strains that do so are sometimes encountered.

Histopathology

Bacteria invade intercellular spaces, where they multiply and produce pectolytic enzymes, including pectin methyl esterase, depolymerase, and pectin lyase. These macerate the tissues by breaking down the middle lamella. Cellulolytic enzymes, produced in much smaller amounts, partially soften the cellulose in the cell walls. Water diffuses from the cell into the intercellular spaces, and the cells collapse and die. Starch is not destroyed except in the later stages of decay.

Disease Cycle

Blackleg. The primary blackleg inoculum is borne on or in seed tubers. After being planted, seed pieces decay at varying times throughout the growing season, releasing large numbers of bacteria into the soil, and sometimes infecting the stem of the host plant (Plate 13). Bacteria may multiply and persist during the growing season in the rhizosphere of the host and, possibly, in the rhizospheres of certain weeds. They may survive the winter in infected stems or tubers. Bacteria will survive in soil for at least a short time. Survival is longer in cool, moist conditions than in warm, dry conditions. The presence of infected plant debris or tubers extends survival of bacteria.

Bacteria may move for some distance in the soil water and contaminate developing daughter tubers of adjacent plants. The amount of contamination of daughter tubers may vary greatly

from season to season depending upon environmental conditions. Bacteria enter lenticels, growth cracks, or injuries at harvest time and can survive in contaminated tubers during the entire storage period. They are readily spread during seed cutting and handling operations.

Soft Rot. Soft rot has a similar disease cycle. Bacteria are also efficiently spread in water used to wash tubers.

Epidemiology

Blackleg. Contamination of seed tubers by *Erwinia* is favored by moist soil and relatively cool temperatures (generally lower than 18–19°C) and is generally more frequent in northern production areas.

Erwinia cells released into the soil from decaying seed pieces survive for varying periods of time, depending upon soil temperature and, to a lesser extent, soil moisture. They may survive for 80–110 days at 2°C but for shorter times at higher temperatures. Some studies have indicated that the half life of *Erwinia* cells in soil is approximately 0.8 days at -29°C, 7.8 days at 0°C, 5.6 days at 7°C, 4.1 days at 13°C, 0.8 days at 18°C, and 0.6 days at 24°C.

Tubers produced under warm (23–25°C or higher), dry conditions are less likely to be contaminated because the pathogens are less likely to survive, and they spread through the soil for shorter distances than when the soil is cool and moist. Tubers harvested from crops grown under dry soil conditions and high soil temperatures may be essentially (or completely) free from *Erwinia* contamination even though they were produced by plants grown from *Erwinia*-contaminated seed tubers.

Cool wet soils at planting time followed by high temperatures after plant emergence favor postemergence blackleg expression; higher soil temperatures favor seed piece decay and pre-emergence death of shoots. Greater total blackleg losses occur in warm areas than in cool ones. *E. carotovora* var. *carotovora* may cause typical blackleg infection if soil temperatures are very high (30–35°C).

Invasion of seed pieces by *Fusarium* spp. may predispose tissues to soft rot and favor blackleg development. High nitrogen fertilization may retard blackleg expression in the field.

Several species of insects disseminate bacteria from potato cull piles or infected plants to seed pieces or stems of healthy growing plants. *Erwinia* aerosols generated by rain or overhead sprinkler irrigation or those generated by mechanical vine destruction may also aid in spreading disease. Mechanical seed piece cutters are responsible for widespread contamination of seed pieces by *Erwinia* spp.

Soft Rot. Soft rot in tubers is favored by immaturity, wounding, solar irradiation, invasion by other pathogens, warm temperatures, high moisture, and lack of oxygen. Tubers harvested at soil temperatures above 20–25°C are highly susceptible. Decay is favored by temperatures above 10°C and retarded by lower temperatures. The optimum temperature for decay by *Erwinia* is above 25–30°C, which is also the optimum range for growth of the pathogens in vitro.

Soft rot caused by species of *Pseudomonas, Bacillus,* and *Clostridium* is favored by temperatures of 30°C or higher.

Anaerobic conditions resulting from poor aeration, flooding of soil, or the presence of a water film on tubers after washing favor disease development. High nitrogen fertilization also increases susceptibility.

Control

Blackleg.
1) Plant seed tubers and, especially, cut seed tubers in well-drained soil.
2) Avoid excessive irrigation to prevent anaerobic soil conditions that favor seed piece decay and subsequent stem invasion.
3) Treat seed tubers with approved fungicides or suberize them well before planting to reduce infection by *Fusarium* spp. and other pathogens that predispose to bacterial invasion.
4) The use of seed tubers derived from stem cuttings may greatly reduce losses caused by blackleg and soft rot. Such seed

should be planted on land with at least two to three years between potato crops, longer if volunteer potatoes are a problem. *Erwinia*-free stocks may be rapidly recontaminated, especially by *E. carotovora* var. *carotovora* under some conditions.

5) Remove potato cull piles, discarded vegetables, and plant refuse to avoid inoculum sources from which insects transmit *Erwinia* spp.

6) Frequently clean and disinfect seed cutting and handling equipment as well as planters, harvesters, and conveyers to eliminate contamination. This should be done at least between different seed lots.

7) Avoid washing seed potatoes unless absolutely necessary, and exercise care during handling operations to reduce damage to seed tubers.

8) Fertilize adequately with nitrogen.

9) To reduce spread of bacteria to healthy plants, remove infected plants as soon as they appear.

Soft Rot

1) Avoid excessive soil moisture before harvest to reduce lenticel infection.

2) Harvest tubers only when mature and only when soil temperatures are less than 20° C. Minimize mechanical damage to tubers during harvesting and handling.

3) Protect harvested tubers from solar irradiation and desiccation.

4) Cool tubers to 10° C or lower as soon as possible after harvest and store at temperatures as low as possible (preferably 1.6–4.5° C). Good ventilation to keep tubers cool and to prevent accumulation of CO_2 and moisture films is especially important.

5) Avoid water films on tuber surfaces, eg, condensation that results from placing tubers with low pulp temperatures into storage with relative humidity above 90%.

6) Do not wash tubers before storage, and when washing them before marketing, dry them as soon as possible and package them in well-aerated containers.

7) Use only clean water to wash potatoes. Contaminated holding tanks used for soaking potatoes almost assure soft rot infection. Treat wash water with chlorine to reduce the amount of soft rot inoculum.

Selected References

BUCHANAN, R. E., and N. E. GIBBONS, eds. 1974. Bergey's Manual of Determinative Bacteriology, 8th ed. Pages 337–338. Williams and Wilkins, Baltimore, MD. 1,268 pp.

CUPPELS, D., and A. KELMAN. 1974. Evaluation of selective media for isolation of soft rot bacteria from soil and plant tissue. Phytopathology 64:468–475.

DeBOER, S. H., and A. KELMAN. 1975. Evaluation of procedures for detection of pectolytic *Erwinia* spp. on potato tubers. Am. Potato J. 52:117–123.

DeLINDO, L., E. R. FRENCH, and A. KELMAN. 1978. *Erwinia* spp. pathogenic to potatoes in Perú. Am. Potato J. 55:383 (Abstr.).

GRAHAM, D. C., and J. L. HARDIE. 1971. Prospects for control of potato blackleg disease by the use of stem cuttings. Proc. Br. Insectic. Fungic. Conf., 6th. pp. 219–224.

HARRISON, M. D., C. E. QUINN, I. A. SELLS, and D. C. GRAHAM. 1977. Waste potato dumps as sources of insects contaminated with soft rot coliform bacteria in relation to recontamination of pathogen-free potato stocks. Potato Res. 20:37–52.

LUND, B. M., and G. M. WYATT. 1972. The effect of oxygen and carbon dioxide concentrations on bacterial soft rot of potatoes. I. King Edward potatoes inoculated with *Erwinia carotovora* var. *atroseptica*. Potato Res. 15:174–179.

MOLINA, J. J., and M. D. HARRISON. 1980. The role of *Erwinia carotovora* in the epidemiology of potato blackleg. II. The effect of soil temperature on disease severity. Am. Potato J. 57:351–369.

NIELSEN, L. W. 1946. Solar heat in relation to bacterial soft rot of early Irish potatoes. Am. Potato J. 23:41–57.

NIELSEN, L. W. 1949. *Fusarium* seedpiece decay of potatoes in Idaho and its relation to blackleg. Idaho Agric. Exp. Stn. Res. Bull. 15. 31 pp.

NIELSEN, L. W. 1978. *Erwinia* species in the lenticels of certified seed potatoes. Am. Potato J. 55:671–676.

O'NEILL, R., and C. LOGAN. 1975. A comparison of various selective media for their efficiency in the diagnosis and enumeration of soft rot coliform bacteria. J. Appl. Bacteriol. 39:139–146.

PEROMBELON, M. C. M. 1974. The role of the seed tuber in the contamination by *Erwinia carotovora* of potato crops in Scotland. Potato Res. 17:187–199.

STEWART, D. J. 1962. A selective-diagnostic medium for the isolation of pectinolytic organisms in the Enterobacteriaceae. Nature 195:1023.

(Prepared by M. D. Harrison and L. W. Nielsen)

Brown Rot

Brown rot, also known as bacterial wilt or southern bacterial wilt, affects potatoes in almost every region in the warm-temperate, semitropical, and tropical zones of the world and has been reported from relatively cool climates. It limits growing of potatoes and other susceptible crops in parts of Asia, Africa, and South and Central America. In the United States the disease occurs in the Southeast from Maryland to Florida. It has rarely occurred in the Southwest or Midwest and has not been confirmed west of the Rocky Mountains.

Symptoms

Field symptoms are wilting, stunting, and yellowing of the foliage. These may appear at any stage in the potato's growth. Wilting of leaves and collapse of stems may be severe in young, succulent plants of highly susceptible varieties. Initially, only one branch in a hill may show wilting. If disease development is rapid, all leaves of plants in a hill may wilt quickly without much change in color. Wilted leaves may fade to a pale green and finally turn brown without rolling of the leaflet edges as they dry (Plate 14). In young potato stems, dark narrow streaks, corresponding to infected vascular strands, become visible through the epidermis.

Brown rot and ring rot have similar but distinguishable symptoms (Table I). A valuable diagnostic sign of brown rot is glistening beads of a gray to brown slimy ooze on the infected xylem in stem cross sections. If the cut surfaces of a sectioned infected stem are placed in contact and then drawn apart slowly, fine strands of bacterial slime become visible and stretch a short distance before breaking.

To demonstrate bacteria in vascular tissue, a longitudinal section from a diseased stem can be placed so that surface tension holds it to the side of a beaker of water and a short segment of the tissue projects below the water surface. Fine milky white strands, composed of masses of bacteria in extracellular slime, stream down from the cut ends of xylem vessels

Table I. Differences Between Brown Rot and Ring Rot

Characteristic	Brown Rot	Ring Rot
Causal organism	*Pseudomonas solanacearum*	*Corynebacterium sepedonicum*
Gram stain	Negative	Positive
Bacterial exudate		
Conditions	Abundant droplets from vascular tissue, usually without squeezing	From vascular tissue, usually with squeezing
Color	Grayish white	Milky white
Symptoms		
Vines	Wilting by rapid collapse; green wilt relatively free from chlorosis	Wilting usually with chlorosis or yellowing; later, necrosis between veins of leaves
Vascular tissue	Distinct browning usually evident in stems	Discoloration in stems, often indistinct
Tubers	Surfaces usually not cracked	Cracks, when present, distributed randomly
Eyes	Exudate causes soil to adhere	Free from adhering soil

29

(Fig. 34B). Bacterial streaming can also be seen microscopically in thin sections of infected tissue mounted in water under a cover slip. This bacterial exudate, combined with wilting and related symptoms, distinguishes this wilt from fungous wilts.

Underground stems, stolons, and roots of plants with initial foliage symptoms show few advanced symptoms of infection. Grayish brown discoloration, usually evident through the tuber periderm, indicates well-established infection. Tubers from infected plants may or may not show symptoms; cross sections usually show distinct, grayish brown vascular discoloration that may extend into the pith or cortex from the xylem tissue. However, certain strains from Portugal and Kenya produce no browning of the vascular ring.

When tubers are cut in half and light pressure is applied, grayish white droplets of bacterial slime ooze out of the vascular ring (Fig. 34A). The eyes, often at the bud or apical end, become grayish brown, and a sticky exudate may form on them or at the stolon connection (Plate 15). The bacterial ooze mixes with the soil, causing soil particles to adhere to the tuber surface. An infected tuber left in the ground continues to decay; secondary organisms convert it to a slimy mass surrounded by a thin layer of cortex and periderm.

Causal Organism

Pseudomonas solanacearum E. F. Smith is a nonspore-forming, noncapsulate, Gram-negative, nitrate-reducing, ammonia-forming aerobic, rod-shaped bacterium. In liquid media, the wild-type bacterium is usually nonmotile and does not form a polar flagellum. Avirulent variants that develop in culture are actively motile (Fig. 34C and D).

Starch is not hydrolyzed by this bacterium, and gelatin is liquefied slowly or not at all. *P. solanacearum* is sensitive to desiccation and is inhibited by relatively low concentrations of salt in broth cultures. Optimal growth of most strains occurs at 30–32° C, although some strains from Colombia grow relatively well at lower temperatures.

Strains differing in biochemical characteristics and host range have been described. A strain pathogenic to potatoes is weakly virulent on tobacco but avirulent on banana; the banana strain is avirulent on potato; in contrast, tobacco and tomato strains are usually virulent on potato. Some strains of *P. solanacearum* from Portugal and Kenya do not form the typical tyrosinase reaction in culture media.

Cultures of *P. solanacearum* maintained in unaerated liquid media rapidly lose virulence and viability and shift from the fluidal (nonmotile) wild-type to avirulent, highly motile variants. Colonies of virulent wild-types are irregularly round and are white with pink centers; colonies of avirulent variants are uniformly round, butyrous, and deep red. Colony characteristics are best observed when cell suspensions are streaked on plates of peptone/casamino acid/glucose agar containing 2,3,5-triphenyltetrazolium chloride and examined in obliquely transmitted light after incubation at 32° C for 36–48 hr.

Epidemiology

In tropical and semitropical regions (southeastern Asia, Central and South America, and parts of Africa and Australia) the pathogen can be borne by tubers; quarantines exist against importation of seed potatoes in some areas. Infected seed potatoes are an important factor in the distribution and increasing severity of the disease in tropical countries such as Peru, where latent infections can occur in seed grown at high elevations.

Temperature plays an important role in the geographic distribution of the organism, which is rare where mean soil temperatures are below 15° C. In North America, seed potatoes are grown in the temperate regions where *P. solanacearum* does not occur, and tuber transmission is not a problem. High temperatures favor growth of the pathogen in vitro and development of the disease in the field. Recently, however, *P. solanacearum* was reported in Sweden and at high altitudes in Costa Rica, Colombia, Peru, and Sri Lanka. Thus, the bacterium can possibly survive and infect potato crops at relatively low temperatures.

The disease occurs in soil types ranging from sandy to heavy clay and over a wide range of soil pH. Disease usually develops in localized areas often associated with poor drainage. On newly cleared forest land, bacterial wilt may be severe if a susceptible crop is planted.

Other Hosts

Important economic hosts of *P. solanacearum* include tobacco, tomato, pepper, eggplant, peanut, banana, and a number of ornamentals and weeds. Although species in over 33 different plant families may be attacked, most susceptible hosts are in the Solanaceae.

Resistance

At least three dominant and independent genes control resistance in potato to certain strains of bacterial wilt. Resistance is relatively sensitive to changes in environment; increased temperatures and decreased light intensity enhance susceptibility to wilt.

Immunity or high levels of resistance have not been identified in clones of *S. tuberosum*. Colombian clones of *S. phureja* Juz. & Buk. with resistance to bacterial wilt have been crossed with haploid lines of *S. tuberosum*. Strains of *P. solanacearum* differing in virulence have complicated breeding for wilt resistance, but two resistant varieties, Caxamarca and Molinera, have been released in Peru.

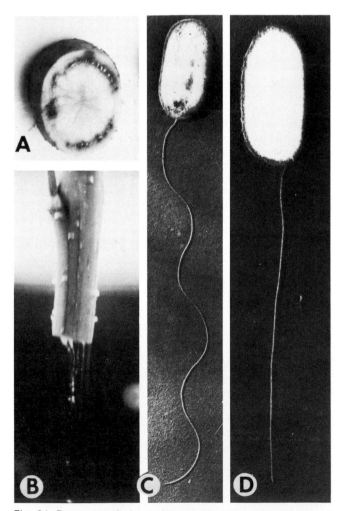

Fig. 34. Brown rot: **A,** bacterial exudate from vascular ring of tuber; **B,** streaming of bacteria from infected stem in water; **C,** cells of *Pseudomonas solanacearum* (electron micrograph) of avirulent type with wavy flagellum; **D,** virulent form (flagellum is atypical). (A, Courtesy L. W. Nielsen; B, courtesy C. Martin; C and D, courtesy A. Kelman)

Control

1) Use disease-free tubers and disinfect the cutting knife.

2) Soil treatment chemicals such as sulfur are not widely accepted because of the low level of control and the high cost.

3) *P. solanacearum* survives for extended periods in some soils. In others it may not survive 1–6 months of fallow.

4) Some crop rotation sequences reduce disease severity; they may act indirectly by reducing populations of root-knot nematodes that enhance bacterial wilt disease infection in potato.

Selected References

BUDDENHAGEN, I., and A. KELMAN. 1964. Biological and physiological aspects of bacterial wilt caused by *Pseudomonas solanacearum.* Annu. Rev. Phytopathol. 2:203–230.

EDDINS, A. H. 1936. Brown rot of Irish potatoes and its control. Fl. Agric. Exp. Stn. Bull. 299. 44 pp.

FELDMESSER, J., and R. W. GOTH. 1970. Association of root knot with bacterial wilt of potato. Phytopathology 60:1014 (Abstr.).

HAYWARD, A. C. 1964. Characteristics of *Pseudomonas solanacearum.* J. Appl. Bacteriol. 27:265–277.

KELMAN, A. 1953. The bacterial wilt caused by *Pseudomonas solanacearum.* N.C. Agric. Exp. Stn. Tech. Bull. 99. 194 pp.

KELMAN, A. 1954. The relationship of pathogenicity in *Pseudomonas solanacearum* to colony appearance on a tetrazolium medium. Phytopathology 44:693–695.

KELMAN, A., and J. HRUSCHKA. 1973. The role of motility and aerotaxis in the selective increase of avirulent bacteria in still broth cultures of *Pseudomonas solanacearum.* J. Gen. Microbiol. 76:177–188.

LOZANO, J. C., and L. SEQUEIRA. 1970. Differentiation of races of *Pseudomonas solanacearum* by a leaf infiltration technique. Phytopathology 60:833–838.

NIELSEN, L. W., and F. L. HAYNES, Jr. 1957. Control of southern bacterial wilt. Potato Handbook Vol. 2. Am. Potato Assoc., New Brunswick, NJ. pp. 47–51.

NIELSEN, L. W., and F. L. HAYNES, Jr. 1960. Resistance in *Solanum tuberosum* to *Pseudomonas solanacearum.* Am. Potato J. 37:260–267.

ROBINSON, R. A. 1968. The concept of vertical and horizontal resistance as illustrated by bacterial wilt of potatoes. Phytopathol. Paper No. 10. Commonw. Mycol. Inst., Kew, Surrey, England. 37 pp.

ROWE, P. R., and L. SEQUEIRA. 1970. Inheritance of resistance to *Pseudomonas solanacearum* in *Solanum phureja.* Phytopathology 60:1499–1501.

STAPP, C. 1965. Die bakterielle Schleimfäule und ihr Erreger *Pseudomonas solanacearum.* Zentralbl. Bakteriol. Parasitenkd. Infektionskr. Hyg. Abt. 2. 119:166–190.

THURSTON, H. D., and J. C. LOZANO. 1968. Resistance to bacterial wilt of potatoes in Colombian clones of *Solanum phureja.* Am. Potato J. 45:51–55.

(Prepared by A. Kelman)

Ring Rot

Ring rot, or bacterial ring rot, was first recorded in Germany in 1906 and has since been found in many other areas. Despite the lack of documentation in a few countries, ring rot has probably occurred wherever potatoes are grown. Through seed certification programs, many countries have successfully eradicated the disease.

Symptoms

Plant symptoms begin with wilting of leaves and stems after midseason. Lower leaves, slightly rolled at the margins and pale green, are usually the first to wilt (Plate 16). As wilting progresses, pale yellowish areas develop between veins. Often only one or two stems of an infected hill develop symptoms. Two important diagnostic features are the wilting of stems and leaves and a milky white exudate that can be squeezed from the vascular ring of tubers (Fig. 35) and of stems when cross-sectioned at their base. A dwarf-rosette type of symptom has been described in the Russet Burbank cultivar in the western United States (Fig. 36).

This disease derives its name from the characteristic internal breakdown in the vascular ring of an infected tuber cross-sectioned at the stem end. Squeezing the tubers, particularly those from storage, expels creamy, cheeselike ribbons of odorless bacterial ooze, which leaves a distinct separation of tissues adjacent to the ring. Secondary invaders (usually soft rot bacteria) cause further tissue breakdown in advanced disease stages, obscuring ring rot symptoms. Pressure developed by this breakdown can cause external swelling, ragged cracks, and reddish brown discoloration, especially near the eyes (Plate 17). Although typical tuber symptoms are invariably apparent in badly infected lots at harvest, some infected tubers may remain symptomless for many weeks in cold storage. Occasionally, typical internal symptoms may not be apparent at the stem end of the tuber but may be found near the apical or rose end.

Causal Organism

Corynebacterium sepedonicum (Spieck. & Kott.) Skapt. & Burkh. is a Gram-positive,[1] nonmotile bacterium. Cells are

[1] Gram stain (Reeds rapid) for bacterial smears: Stains are Gram-positive bacterial cells blue and Gram-negative cells pink. Mix in equal parts: 1) crystal or gentian violet 0.25% aqueous with 2) NaHCO₃ 1.25% aqueous. Flood for 10 sec and drain. Mix 20% iodine in 1M NaOH and dilute 1:10 in water. Flood for 10 sec and wash. Mix 1 part acetone and 3 parts 95% ethyl alcohol. Rinse until no more color washes from smear. Flood with water. Dilute basic fuchsin saturated in ethyl alcohol 1:10 in water. Stain not over 2 sec, rinse thoroughly, and dry. (From Glick, D. P., P. A. Ark, and H. N. Racciot. 1944. Am. Potato J. 21:311–314.)

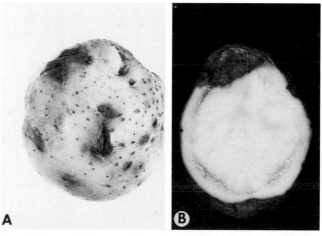

Fig. 35. Bacterial ring rot: **A,** surface cracking from *Corynebacterium sepedonicum* infection; **B,** cheesy breakdown of tuber vascular tissue.

Fig. 36. Dwarf rosette symptom of ring rot. (Courtesy J. R. Letal)

0.4–0.6 × 0.8–1.2 μm and predominantly wedge-shaped, although curved and straight rods are also present. Single cells are most abundant, but V and Y configurations are often observed. Growth on all media is slow, and colonies rarely exceed 1 mm in diameter after five days on nutrient glucose agar.

Disease Cycle

The organism overwinters primarily in infected tubers, either those in storage or those that survive the winter in the field. It apparently cannot survive in unsterilized field soil, although it remains viable for nine months or more in dried slime on bags, crates, etc., in storage and in unprotected sites. Infection occurs through tuber wounds, especially from contaminated machinery and containers. Contaminated seed-cutting knives and picker planters are excellent disseminators. Invasion also occurs through wounds in stems, roots, stolons, or other plant parts, and transmission has been experimentally obtained through tomato seed. Root inoculation is highly efficient and hastens symptom development. Bacteria become established in large vessels and later invade xylem parenchyma and adjacent tissue and cause separation at the vascular ring. Certain sucking insects transmit the disease from diseased to healthy plants.

Epidemiology

Conditions for dissemination of the pathogen are most favorable in the spring when infected seed tubers are warmed before planting, thus increasing bacterial activity. Surfaces of freshly cut seed provide ideal infection courts. Disease develops most rapidly at 18–22° C soil temperatures, but higher temperatures decrease infection from seed piece inoculation. In general, warm, dry weather hastens symptom development, but temperatures above optimum delay symptom expression.

Disease-resistant cultivars have been developed, but immunity is as yet unknown. No resistant cultivar has achieved economic prominence, and such cultivars may still serve as carriers.

Other Hosts

Only *S. tuberosum* is affected naturally, although 28 *Solanum* spp. and two *Lycopersicon* spp. have been demonstrated experimentally to be symptom-producing hosts. Inoculation assays with young eggplants or tomato plants demonstrate the presence of ring rot bacteria in suspected potato plants.

Control

Use of disease-free seed accompanied by strict sanitation procedures is the only method of control.
1) Dispose of all potatoes from the farm when the disease is found.
2) Thoroughly disinfect warehouses, crates, and handling, planting, harvesting, and grading machinery.
3) Use new bags for clean seed because disinfection of bags is not effective.
4) Plant seed that is free of ring rot. Seed certification programs regularly reject stocks in which any ring rot is found.
5) Do not plant disease-free seed in a field with volunteer plants from a previously infected crop.

Selected References

BONDE, R., and M. COVELL. 1950. Effect of host variety and other factors on pathogenicity of potato ring-rot bacteria. Phytopathology 40:161–172.
CLAFLIN, L. E., and J. F. SHEPARD. 1977. An agglutination test for the serodiagnosis of *Corynebacterium sepedonicum*. Am. Potato J. 54:331–338.
DeBOER, S. H., and R. J. COPEMAN. 1974. Endophytic bacterial flora in *Solanum tuberosum* and its significance in bacterial ring rot diagnosis. Can. J. Plant Sci. 54:115–122.
DUNCAN, J., and H. GENEREUX. 1960. La transmission par les insectes de *Corynebacterium sepedonicum* (Spieck. and Kott.)
Skaptason and Burkholder. Can. J. Plant Sci. 40:110–116.
GUTHRIE, J. W. 1959. The early, dwarf symptom of bacterial ring rot of potato in Idaho. Phytopathology 49:453–454.
KNORR, L. C. 1948. Suscept range of the potato ring rot bacterium. Am. Potato J. 25:361–371.
LELLIOTT, R. A., and P. W. SELLAR. 1976. The detection of latent ring rot (*Corynebacterium sepedonicum* (Spieck. et Kotth.) Skapt. et Burkh.) in potato stocks. Eur. and Mediterr. Plant Prot. Org. (EPPO) Bull. 6:101–106.
PAQUIN, R., and H. GENEREUX. 1976. Effet du climat sur la fletrissure bacterienne de la pomme de terre et relation avec le contenu en sucres des tiges. Can. J. Plant Sci. 56:549–554.
REPORT TO CERTIFICATION COMMITTEE OF THE POTATO ASSOCIATION OF AMERICA. 1957. How can we interpret the zero tolerance for bacterial ring rot in certified seed potatoes? Am. Potato J. 34:142–148.
SLACK, S. A., H. A. SANFORD, and F. E. MANZER. 1979. The latex agglutination test as a rapid serological assay for *Corynebacterium sepedonicum*. Am. Potato J. 56:441–446.

(Prepared by F. Manzer and H. Genereux)

Pink Eye

The disease is of minor importance, and little is known about factors influencing tuber infection and disease development. It is not a problem with certain cultivars, but in susceptible cultivars incidence may be high. Pink eye is frequent in tubers of plants infected with Verticillium wilt.

Symptoms

Pink areas around the eye later turn brown. Discolored areas are particularly abundant at the apex of the tuber and are usually superficial but may extend into the tuber 8 mm or more. Internal discoloration may result in cavities.

Symptoms are most conspicuous at harvest, particularly following high soil moisture levels during tuber formation (Plate 18). In dry storage, tissue superficially affected with pink eye soon dries out, becoming scalelike and inconspicuous. In storage at high relative humidity, especially at high temperatures, rot may follow pink eye (Plate 19). Infection follows bruising and is involved in seed tuber decay.

Although the disease is associated with Verticillium wilt, both diseases occur independently. Infection has also been linked to *Rhizoctonia* and to the wet rot phase of late blight infection. The color of affected tissue is very similar to that from late blight.

The red xylem symptom follows infection of the stolon end, causing a scar at the stolon attachment or reddish brown vascular discoloration.

Causal Organism

The pathogen was identified in early investigations as *Pseudomonas fluorescens* Migula Gram-negative rods, 0.3–0.4 × 1.0–1.3 μm in size, occurring both singly and in pairs. Cells are usually motile and possess a polar flagellum but are occasionally nonmotile. Pectolytic enzymes are apparently active in pathogenesis. This identification of the causal bacterium is presently in considerable doubt. Much more information is needed on types of bacteria present in tuber lesions, pathogenicity of isolates to potato tubers, and identification of the pathogen or pathogens involved.

Control

Store tubers under cool dry conditions to dry out affected tissue and to prevent disease progression in storage.

Selected References

CUPPELS, D. A., and A. KELMAN. 1980. Isolation of pectolytic fluorescent pseudomonads from soil and potatoes. Phytopathology 70:1110–1115.
DOWSON, W. J., and D. R. JONES. 1951. Bacterial wet rot of potato tubers following *Phytophthora infestans*. Ann. Appl. Biol. 38:231–236.

FOLSOM, D., and B. A. FRIEDMAN. 1959. *Pseudomonas fluorescens* in relation to certain diseases of potato tubers in Maine. Am. Potato J. 36:90–97.

FRANK, J. A., R. E. WEBB, and D. R. WILSON. 1973. The relationship between Verticillium wilt and the pinkeye disease of potatoes. Am. Potato J. 50:431–438.

SANDS, D. C., and L. HANKIN. 1975. Ecology and physiology of fluorescent pectolytic pseudomonads. Phytopathology 65:921–924.

(Prepared by W. J. Hooker)

Bacteria in Potatoes That Appear Healthy

Bacterial populations of various types are frequently present in stems and tubers of apparently healthy plants. *Bacillus megaterium* de Bary is common, as are some apparently non-pathogenic Gram-positive bacteria and strains of *Micrococcus, Pseudomonas, Xanthomonas, Agrobacterium,* and *Flavobacterium.* Populations are high in vascular tissue, possibly resulting from root invasion, and are present in higher numbers in plants from cut seed than in those from whole seed. The role of these apparently nonpathogenic organisms is unknown. Frequency of reports and differences in types of organisms observed suggest that considerable variation exists.

Selected References

CIAMPI, L. R., H. J. DUBIN, and C. JOFRE. 1976. La flora bacteriana vascular en tubérculos de papa en el sur de Chile. Fitopatología 11:57–61.

DeBOER, S. H., and R. J. COPEMAN. 1974. Endophytic bacterial flora in *Solanum tuberosum* and its significance in bacterial ring rot diagnosis. Can. J. Plant Sci. 54:115–122.

HOLLIS, J. P. 1951. Bacteria in healthy potato tissue. Phytopathology 41:350–366.

(Prepared by W. J. Hooker)

Common Scab

This disease, present to some extent in most areas where potatoes are grown, is a major production problem that affects grade quality and has only a small effect on total yield or storing ability.

Symptoms

Tuber lesions are usually circular, 5–8 (seldom exceeding 10) mm in diameter, but they may be irregular in shape and larger when infections coalesce. Affected tissues vary from light tan to brown. They may consist of a superficial corklike layer (russet scab), an erumpent or cushionlike scab (raised scab) 1–2 mm high, or an extension into the tuber (pitted scab) of various depths to 7 mm (Plate 20). Pitted lesions are dark brown or almost black (Plate 21). Tissue under the lesion is straw color and somewhat translucent; under russet scab such tissue may not be evident.

Brown to tan stem and stolon lesions originate at lenticels as elongate lens-shaped lesions or at other natural wounds (emergence points of roots or split portions of stems) as approximately circular lesions.

Naturally occurring aboveground symptoms have not been reported, but leaves of potato and other plants have been infected experimentally.

Causal Organism

Streptomyces scabies (Thaxter) Waksman & Henrici (syn. *Actinomyces scabies* (Thaxter) Güssow) has barrel-shaped conidia 0.8–1.7 × 0.5–0.8 μm. Conidiophores are branched, having septa, of the "attenuated isthmus" type with long, spirally coiled, terminal branches (Fig. 37). Other *Streptomyces* spp. of lower virulence have been described as pathogenic, but more work is needed to properly evaluate differences in pathogenicity among species. An acid-tolerant pathogenic species distinct from *S. scabies* and as yet unnamed can survive in soils with pH below 5.2.

Streptomycetes are classified with bacteria because they are

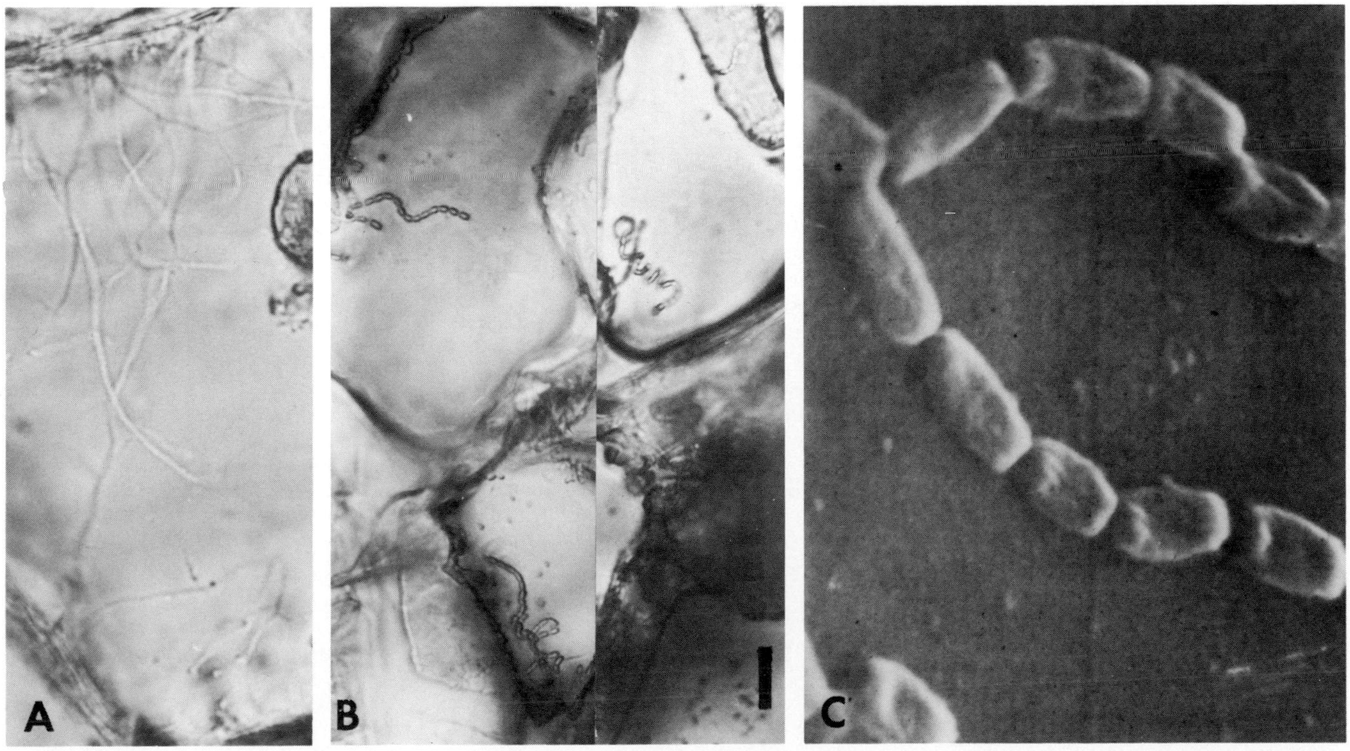

Fig. 37. *Streptomyces scabies:* **A,** vegetative filaments within tuber tissue infected with common scab; **B,** coils of spores and vegetative filaments within cells near surface (bar represents 10 μm); **C,** spores photographed by scanning electron microscope (×10,000). (C, Courtesy G. A. McIntyre)

akaryotic and have cell wall biochemical characteristics more closely resembling those of bacteria than of fungi. They do resemble fungi in their filamentous morphology but differ notably in the small (approximately 1-μm) diameter of their vegetative filaments.

S. scabies is aerobic. It produces colorless vegetative filaments and pale, mouse-gray aerial mycelia on a number of media, often with melanin pigmentation of the medium surrounding the colony. Sporulation is good on potato agar media with low sucrose (0.5%) levels and is sparse to lacking on media rich in peptone. *S. scabies* can usually be successfully isolated from the straw-colored translucent tissue immediately below the lesion by dilution plating with soil-water agar or low (0.1–0.5%) sucrose potato agar. Light reflectance from *Streptomyces* colonies is distinct from that of typical bacterial colonies because of the radiating *Streptomyces* filaments. Growth in culture ranges from 5 to 40.5° C; optimum temperature is 25–30° C.

Other Hosts

The organism causes scab on the fleshy roots of other plants such as beets (red and sugar), radish, rutabaga, turnip, carrot, and parsnip. In these, it is seldom of economic importance. Fibrous roots of potato and other plants are also susceptible.

Histopathology

Actively growing tubers are infected through young lenticels and probably also through stomata of the epidermis before the periderm differentiates. Portions of tubers protected by well-developed periderm are not susceptible. Wounds also serve as infection sites. Insect larval feeding aids initial penetration and progression through wound periderm layers. Filaments are small and difficult to detect in tissue. They are believed to be initially intercellular and to become intracellular as tissue involvement progresses.

Scab lesions may be deep or shallow. The type of lesion is frequently determined genetically by potato resistance or susceptibility. Resistance is apparently associated with the effectiveness of the periderm, which underlies lesions and walls them off from the tuber. In susceptible cultivars with deep scab lesions, successive periderm layers form as penetration progresses. In resistant cultivars, lesions are shallow and a single periderm layer seems to prevent further infection. Different lesion types may occur on the same tuber, which may reflect differences in pathogenicity between infection propagules or in maturity of the tuber surface at the time of infection.

Epidemiology

S. scabies has been introduced into virtually all potato soils by infected potato seed. Evidence exists, however, that pathogenic streptomycetes were present in native soils before potatoes were introduced. The organism is essentially a low-grade saprophytic pathogen that survives for long periods on decaying plant parts in the soil or possibly on roots of living plants, in old feed lots, or in fields heavily manured with animal wastes.

Physiological specialization of parasite subculture isolates to different potato genotypes have been demonstrated in greenhouse trials. In the field, however, selective pathogenicity by biotypes is usually of minor importance because the resistance or susceptibility of potato cultivars remains relatively constant over a wide range of natural soil populations.

Continuous crops of potatoes generally increase severity of scab. In contrast, as time between successive crops is increased, scab severity decreases to a relatively constant level.

Maintaining adequate soil moisture during tuber set and enlargement is critical in controlling the extent of scab. Field irrigation after tuber set and during enlargement reduces scab appreciably. Optimum levels of soil moisture are those at field capacity, which favor optimum potato growth. Permitting tubers to develop in infested dry soil increases incidence of scab in susceptible genotypes.

Sulfur has been applied to reduce soil pH. However, reduction of scab cannot regularly be explained as an effect of soil pH alone. Manganase applications have reduced scab, and the reduction of the calcium-phosphorus ratio of the soil has reduced scab severity.

Chemical treatment of soil depends on the proper incorporation of chemicals into the soil and should be coordinated with other prevention methods. Ideally, chemical treatments should be effective in subsequent years and increased crop value should justify costs. Pentachloronitrobenzene has been widely tested. When it is properly mixed into the soil, results are generally beneficial the first year; residual effects are less or nonexistent in following years. Repeated application may be necessary and treatment costs may be high. Band application is more economical than broadcast, and small yield increases are possible. Urea formaldehyde liquid as a furrow application has been successful in certain trials.

Control

Prevention depends on a combination of practices.

1) Avoid planting scabby seed tubers.

2) Increase time between successive potato crops. This reduces scab incidence to a relatively constant level but seldom completely eliminates soil populations.

3) Varietal resistance or susceptibility determines incidence and lesion type. Cultivars with low levels of resistance nevertheless often produce a marketable crop in the presence of scab.

4) Maintain high soil moisture levels during and after tuber set for 4–9 weeks as determined by cultivar, growing practices, and climate.

5) Avoid overliming of soil, which increases soil pH and lowers soil Ca-P ratio.

6) Soil treatments include: sulfur and acid-forming fertilizers to increase soil acidity, pentachloronitrobenezene, urea formaldehyde, and other soil fumigants.

7) Seed treatments with organic mercury avoid introducing inoculum to new areas but are not permitted in certain countries. Effectiveness of other treatments in destroying tuberborne inoculum has not been established.

8) Mancozeb (8%) dust as a tuber seed treatment effectively controls acid scab.

Selected References

DAVIS, J. R., G. M. McMASTERS, R. H. CALLIHAN, F. H. NISSLEY, and J. J. PAVEK. 1976. Influence of soil moisture and fungicide treatments on common scab and mineral content of potatoes. Phytopathology 66:228–233.

DAVIS, J. R., R. E. McDOLE, and R. H CALLIHAN. 1976. Fertilizer effects on common scab of potato and the relation of calcium and phosphate-phosphorus. Phytopathology 66:1236–1241.

HOFFMAN, G. M. 1959. Untersuchungen zur physiologischen spezialisierung von *Streptomyces scabies* (Thaxt.) Waksman et Henrici. Zentralbl. Bakteriol. Parasitenkd. Infectionskr. Hyg. Abt. 2. 112:369–381.

HOOKER, W. J., and O. T. PAGE. 1960. Relation of potato tuber growth and skin maturity to infection by common scab, *Streptomyces scabies*. Am. Potato J. 37:414–423.

JONES, A. P. 1965. The streptomycetes associated with common scab of the potato. Plant Pathol. 14:86–88.

LAPWOOD, D. H., L. W. WELLINGS, and J. H. HAWKINS. 1973. Irrigation as a practical means to control potato common scab (*Streptomyces scabies*): Final experiment and conclusions. Plant Pathol. 22:35–41.

MANZER, F. E., G. A. McINTYRE, and D. C. MERRIAM. 1977. A new potato scab problem in Maine. Maine Agric. Exp. Stn. Tech. Bull. 85. 24 pp.

STAPP, C. 1956. Streptomykose der Kartoffeln (Kartoffelschorf). Pages 494–534 in: O. Appel and H. Richter, eds. Bakterielle Krankheiten. Handb. Pflanzenkr. 2. Paul Parey: Berlin-Hamburg.

(Prepared by W. J. Hooker)

Fungi

Powdery Scab

Although powdery scab develops best under cool, moist conditions, it is found in practically every potato-producing area in the world from latitudes 65° N to 53° S and at higher altitudes in the tropics.

Symptoms

Tuber infection in lenticels, wounds, and (less frequently) in the eyes is evident as purplish-brown pustules, 0.5–2 mm in diameter, extending laterally under the periderm and forming a raised or pimplelike lesion. Enlargement and division of host cells force the periderm to rupture, resulting in white, wartlike outgrowths (Plate 22).

Wound periderm forms beneath the lesion, which gradually darkens and decays, leaving a shallow depression filled with a powdery mass of dark brown spore balls (cystosori) (Fig. 38, Plate 22). The lesion is usually surrounded by the raised, torn edges of the burst periderm. If, in very wet soil, wound periderm does not develop, the lesion expands in depth and width, forming hollowed-out areas or very large warts. This is the cankerous form of powdery scab.

In storage, powdery scab may lead to a dry rot or to more warts or cankers. If infected tissue has not burst through the periderm, infection and necrosis may spread laterally, producing one or two necrotic rings surrounding the original infection. Under humid conditions, after the periderm has ruptured, warts may become somewhat larger and secondary warts may develop beside the primary warts with little or no necrosis beneath the skin.

Powdery scab lesions may serve as infection courts for late blight and a number of wound pathogens.

Infection on roots and stolons parallels that on tubers, with small necrotic spots developing into milky white galls varying in diameter from 1 to 10 mm or more. Galls on roots may become so severe that young plants wilt and die. As galls mature, they turn dark brown (Plate 23) and gradually break down, liberating powdery masses of cystosori into the soil. Galls superficially resemble those of wart, except that *S. endobioticum* does not attack roots.

Causal Organism

Spongospora subterranea (Wallr.) Lagerh. f. sp. *subterranea* Tomlinson is a member of the Plasmodiophorales. Cystosori are ovoid, irregular, or elongate, 19–85 μm in diameter, and consist of an aggregate of closely associated resting spores (cysts). Each spore is polyhedral, 3.5–4.5 μm in diameter, with smooth, thin, yellow-brown walls. Primary and secondary zoospores are uninucleate, ovoid to spherical, 2.5–4.6 μm in diameter, with two flagella of unequal length (e.g., 13.7 and 4.35 μm).

Disease Cycle

The fungus survives in soil in the form of cystosori made up of resting spores. Stimulated by the presence of roots from susceptible plants, resting spores germinate to produce primary zoospores. These penetrate epidermal cells of roots and stolons or root hairs, ultimately producing multinucleate fungus masses (sporangial plasmodia), which yield secondary zoospores that further spread infection to roots and tubers. Invasion by secondary zoospores stimulates the host cells to become larger and more numerous, and galls are produced. Within these galls, balls of resting spores are ultimately formed (Fig. 39).

Epidemiology

Inoculum is spread by soil and by tuberborne resting spores. Tuber and root infection is favored by cool, moist soil conditions in the earlier stages of infection and later by gradual drying of the soil. Cysts may persist in the soil for up to six years.

The time from tuber and root infection to gall formation is less than three weeks at a temperature of 16–20° C. Powdery scab occurs in field soils ranging from pH 4.7 to 7.6.

Fertilization experiments with N, P, K, ammonium sulfate, calcium nitrate, and minor elements have shown generally that nutrition of the soil has little or no effect on incidence of powdery scab. However, sulphur added to the soil can decrease the intensity of scabbing. Recent studies indicate that zinc oxide incorporated into soil reduces the amount of scab.

The effect of liming is not clear. In some areas, liming of the soil has resulted in an increase of powdery scab, whereas in others, liming has decreased or had no effect on incidence of the disease.

Spores survive passage through the digestive tracts of animals.

S. subterranea is a vector for potato mop-top virus.

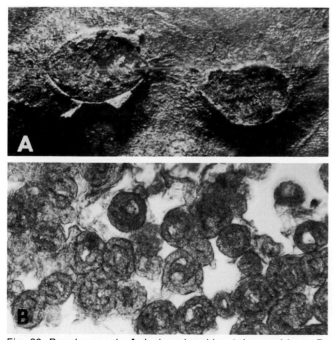

Fig. 38. Powdery scab: **A**, lesions breaking tuber periderm; **B**, spore balls of *Spongospora subterranea* within lesion. (A, Courtesy R. Salzmann and E. R. Keller; B, courtesy C. H. Lawrence)

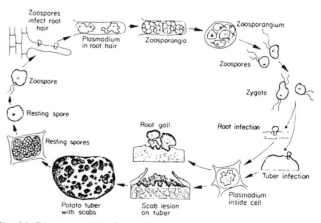

Fig. 39. Disease cycle of powdery scab caused by *Spongospora subterranea*. (Courtesy G. N. Agrios)

Therefore, control can concomitantly diminish incidence of mop-top.

Other Hosts

The fungus infects and completes its life cycle on other tuber-bearing *Solanum* spp. and on roots of nontuber-bearing *S. nigrum* L. and *Nicotiana rustica* L. Other hosts that are infected without formation of resting spores include dicotyledons, monocotyledons, and a gymnosperm.

Control

1) No completely adequate measures have been developed. Resistant cultivars are recommended, but no immune varieties are known.

2) Crop rotations of three to 10 years have been recommended, depending on climate and soil conditions.

3) Plant disease-free seed.

4) Crop in porous and well-drained soils, and avoid planting on land known to be contaminated.

5) Do not use manure from animals fed infected tubers.

6) Fertilizers and other chemical soil treatments are generally not effective. Sulphur has given beneficial results but its use is limited because soil may be made too acid for optimum potato growth.

7) Soaking infected seed tubers in solutions of formaldehyde or mercuric chloride reduces seedborne inoculum.

Selected References

COOPER, J. I., R. A. C. JONES, and B. D. HARRISON. 1976. Field and glasshouse experiments on the control of potato mop-top virus. Ann. Appl. Biol. 83:215–230.

HIMS, M. J., and T. F. PREECE. 1975., *Spongospora subterranea*. No. 477 in: Descriptions of Pathogenic Fungi and Bacteria. Commonw. Mycol. Inst., Kew, Surrey, England. 2 pp.

KARLING, J. S. 1968. Powdery Scab of Potatoes, and Crook Root of Watercress. Pages 180–192 in: J. S. Karling, ed. The Plasmodiophorales. Hafner Pub. Co., New York and London.

KOLE, A. P. 1954. A contribution to the knowledge of *Spongospora subterranea* (Wallr.) Langerh., the cause of powdery scab of potatoes. Tijdschr. Plantenziekten 60:1–65.

KOLE, A. P., and A. J. GIELINK. 1963. The significance of the zoosporangial stage in the life cycle of the Plasmodiophorales. Neth. J. Plant Pathol. 69:258–262.

WENZL, H. 1975. Die Bekämpfung des Kartoffelschorfes durch Kulturmassnahmen. Z. Pflanzenkr. Pflanzenschutz 82:410–440.

(Prepared by C. H. Lawrence and A. R. McKenzie)

Wart

Wart has been recorded in Africa, Asia, Europe, and North and South America. In certain areas, disease spread has been contained through strict quarantine.

Symptoms

Warty outgrowths or tumorous galls, pea-sized to the size of a man's fist, develop at the base of the stem. Aboveground galls are green to brown, becoming black at maturity and later decaying. Occasionally galls form on the upper stem, leaf, or flower. Belowground galls appear at stem bases, stolon tips, and tuber eyes (Plate 24). Tubers may be disfigured or completely replaced by galls (Plate 25). Subterranean galls are white to brown, becoming black through decay. Roots are not known to be attacked.

Causal Organism

Synchytrium endobioticum (Schilb.) Perc. does not produce hyphae but enters the host epidermis as a zoospore, swells to a prosorus, and develops into a sorus. Haploid sori form inside the cells, each sorus containing 1–9 sporangia. Resting or winter sporangia are golden brown, spheroidal, measuring 35–80 μm in diameter. The thick sporangium wall is prominently ridged, generally with three ridges confluent at two sides of the sporangium. Zoospores measure 1.5–2.2 μm in diameter, are pear-shaped, and are motile by a single posterior flagellum.

The fungus exists in a number of physiologic races, making evaluation of resistance difficult.

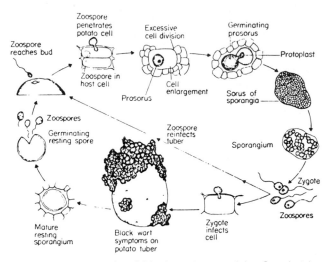

Fig. 40. Disease cycle of black wart caused by *Synchytrium endobioticum*. (Reprinted, by special permission, from Plant Pathology, 2nd ed., by G. N. Agrios. ©1978 The Academic Press, New York)

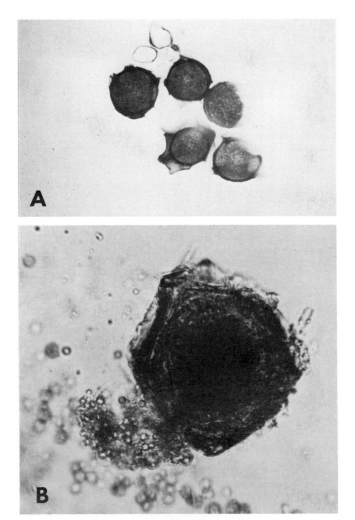

Fig. 41. *Synchytrium endobioticum* winter sporangia (**A**, resting; **B**, germinating) range in diameter from 35 to 80 μm. (A, Courtesy M. E. Hampson; B, courtesy F. Frey)

Histopathology

Sori of sporangia develop in epidermal cells of meristematic tissues of growing points, buds, stolon tips, or young leaf primordia. Invaded and surrounding cells enlarge. Rapid cell division following infection from either zygotes or haploid zoospores causes an increase in meristematic tissue, providing additional infection courts. The gall is a malformed branch system largely composed of thin-walled parenchyma. In near-immune cultivars, the warts remain superficial and scablike, whereas, in resistant cultivars, the zoospore dies soon after invasion by necrotic abortion (hypersensitive reaction) of the infected tissue.

Disease Cycle

From the initial infection, summer sporangia form as membranous sacs, producing motile zoospores (Fig. 40). Zoospores encyst and penetrate epidermal cells of susceptible tissue approximately 2 hr after formation. After a developmental period, zoospores are released outside the plant. These may reinfect surrounding meristematic tissue in a secondary cycle, or they may conjugate forming a zygote that reinfects several cells deep, giving rise to the resting or winter sporangium (Fig. 41). Resting sporangia are released from galls through decay of host tissue. The fungus can survive in soil as resting sporangia for as long as 38 years, even through adverse conditions. Spread of inoculum is through soil-infested tubers, implements, containers, etc. Resting sporangia germinate erratically, producing zoospores.

Epidemiology

The fungus is most active where susceptible tissue is present in growing sprouts, stolons, buds, and eyes. Water is required for germination of winter and summer sporangia and for zoospore distribution. Cool summers with average temperature of 18° C or less, winters of approximately 160 days of 5° C or less, and an annual precipitation of 70 cm are necessary for disease development, and disease is limited to such environments. Soil reaction is of less importance, with disease occurring in soil with pH from 3.9 to 9.5. Temperatures of 12–24° C favor infection.

S. endobioticum is reported to serve as vector for potato virus X.

Other Hosts

Potato is the principal host, although *S. endobioticum* has been experimentally transferred to a number of the Solanaceae. Roots and leaves are attacked, with resting sporangia developing sparsely. Tomato cultivars are particularly susceptible. Plants other than potato are not believed to be of importance in the disease cycle. Some South American wild potato species show resistance to a variable extent.

Control

1) Worldwide control of spread is being attempted through quarantine legislation.
2) Resistant cultivars have been developed in Europe and North America.
3) No chemical control is known that is not also injurious to soil and crops.
4) The herbicide dinoseb (2-sec-butyl-4,6-dinitrophenol) reduces infection to some extent.

Selected References

BRAUN, H. 1959. Rassenbildung der Kartoffelkrebs (*Synchytrium endobioticum*). Kartoffelbau 10:234–237.
CURTIS, K. M. 1921. The life-history and cytology of *Synchytrium endobioticum* (Schilb.) Per., the cause of wart disease in potato. Philos. Trans. R. Soc. London Ser. B 210:409–478.
HAMPSON, M. C., and K. G. PROUDFOOT. 1974. Potato wart disease, its introduction to North America, distribution and control problems in Newfoundland. FAO Plant Prot. Bull. 22:53–64.
KARLING, J. S. 1964. *Synchytrium*. Academic Press, New York. 470 pp.
NOBLE, M., and M. D. GLYNNE. 1970. Wart Disease of Potatoes. FAO Plant Prot. Bull. 18:125–135.

(Prepared by M. C. Hampson)

Skin Spot

The disease is common in northern Europe and also occurs in North America and Australasia.

Symptoms

Following infection, roots, stolons, and stems belowground develop discrete, light brown lesions that enlarge, darken, and crack transversely. Cortical tissues may become detached. In storage, purplish-black, slightly raised spots up to 2 mm in diameter form on tubers singly or in groups, either at random over the surface or aggregated around eyes, stolon scars, or damaged skin (Fig. 42A). Sometimes larger necrotic areas form over the tuber surface. Brown lesions may occur on sprouts of tubers stored in humid conditions (Fig. 42B). Irregular stands and delayed plant emergence frequently follow severe infection of seed tubers.

Causal Organism

Polyscytalum pustulans (Owen & Wakef.) M. B. Ellis (syn. *Oospora pustulans* Owen & Wakef.) has erect, branched conidiophores up to 140 μm long, the lower part pale brown and sometimes swollen at the base (Fig. 42C). Conidia are dry, cylindrical, mostly unicellular but occasionally 1-septate, measuring $6-18 \times 2-3$ μm. They develop in chains that fragment readily. In culture, colonies are gray and powdery. Sclerotia up to 1 mm in diameter may form in aging cultures.

Histopathology

Conidia germinate and infect tubers through lenticels, eyes, and skin abrasions. The fungus penetrates tuber tissue to a depth of about 12 cells and within two weeks is checked by formation of a cork cambium; after two months, infected tissues degenerate and spots become visible. If formation of a cambium is delayed or prevented, e.g., by treatment with certain sprout inhibitors, the fungus can penetrate deeper into tuber tissue, subsequently causing larger spots to develop. Toward the end of the growing season, sclerotia form within cells of decaying cortical tissues of stems and tubers.

Disease Cycle

First infections originate from inoculum on seed tubers. Brown lesions develop initially on stems near the attachment to the seed tuber and later at nodes. Stolons are often first infected at the apex, and the fungus infects buds and bud scales of tuber eyes as they form. Further infection of underground parts before harvest and of tubers during harvest and in storage originates from conidia produced abundantly on infected tissues. Infected tubers usually appear symptomless when harvested; skin spots and necrotic buds in eyes develop about two months later and increase in storage.

Epidemiology

The disease is prevalent in cool wet seasons and is more severe in crops grown on heavy clay loam than in those grown on sandy or organic soils. Skin spot is common in crops lifted moist and stored cool (4° C), and conidia produced on the spots spread infection in storage. Cultivars with thick tuber skin develop few skin spots. Cultivars susceptible to skin infection are usually also susceptible at tuber eyes. Most inoculum originates from infected seed tubers, but sclerotia of *P. pustulans* in decaying stems can remain viable in soil for at least eight years.

Other Hosts

Skin spot develops only on potatoes, but brown lesions can develop on roots of other Solanaceae (*Solanum* spp., *Lycopersicon esculentum*, *Nicotiana* spp., and *Datura* spp.).

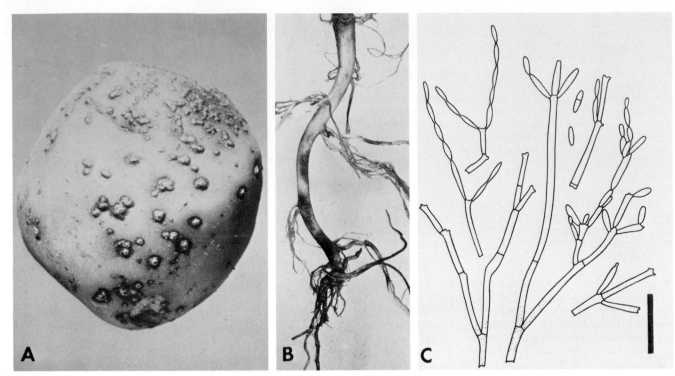

Fig. 42. Skin spot: **A,** on potato tuber; **B,** on stem and roots; **C,** *Polyscytalum pustulans (Oospora pustulans).* Bar represents 25 μm. (A, Copyright National Institute of Agricultural Botany, Cambridge, England; B, courtesy Rothamsted Experimental Station, Harpenden, Herts, England; C, reprinted, by special permission, from More Dematiaceous Hypomycetes, by M. B. Ellis. ©1976 Commonwealth Mycological Institute, Kew, Surrey, England)

Control

1) Skin spots and damage to tuber eyes can be prevented by storing tubers dry (75% rh) and warm (15°C), although the fungus remains viable in the infections.

2) Disinfection of seed tubers soon after harvest with fungicides, including organomercurials and 2-aminobutane, is effective in preventing disease during storage.

3) Benomyl and thiabendozole applied at harvest also decrease the disease. Because these materials persist in the tuber skin and prevent sporulation of *P. pustulans* on seed tubers, they greatly reduce subsequent infection of plants and progeny tubers.

4) Tubers free from *P. pustulans* can be produced by propagating plants from stem cuttings, although, to maintain the health of stocks, treatment with fungicide is needed to prevent reinfection during their commercial multiplication.

Selected References

BOYD, A. E. W. 1972. Potato storage diseases. Rev. Plant Pathol. 51:297–321.

BOYD, A. E. W., and J. H. LENNARD. 1962. Seasonal fluctuation in potato skin spot. Plant Pathol. 11:161–166.

GRAHAM, D. C., G. A. HAMILTON, C. E. QUINN, and A. D. RUTHVEN. 1973. Use of 2-aminobutane as a fumigant for control of gangrene, skin spot and silver scurf diseases of potato tubers. Potato Res. 16:109–125.

HIDE, G. A., J. M. HIRST, and E. J. MUNDY. 1969. The phenology of skin spot (*Oospora pustulans* Owen and Wakef.) and other fungal diseases of potato tubers. Ann. Appl. Biol. 64:266–279.

HIDE, G. A., J. M. HIRST, and O. J. STEDMAN. 1973. Effects of skin spot (*Oospora pustulans*) on potatoes. Ann. Appl. Biol. 73:151–162.

HIRST, J. M., G. A. HIDE, R. L. GRIFFIN, and O. J. STEDMAN. 1970. Improving the health of seed potatoes. J. Agric. Soc. Engl. 131:87–106.

HIRST, J. M., and G. A. SALT. 1959. *Oospora pustulans* Owen and Wakefield as a parasite of potato root systems. Trans. Br. Mycol. Soc. 42:59–66.

KHARAKOVA, A. P. 1961. On the biology of the causal agent of oosporosis, *Oospora pustulans* Owen and Wakefield. Rev. Appl. Mycol. 40:558.

(Prepared by G. A. Hide)

Leak

Leak, also called watery wound rot, may occur sporadically wherever potatoes are grown.

Symptoms

Only tubers are affected. A discolored, water-soaked area appears around a bruise or cut on the skin. As disease develops, the tuber appears to be swollen and the skin is moist. Internally diseased flesh is clearly demarcated from healthy tissue by a dark boundary line. Rotted tissue is spongy, wet, and may have cavities. On cutting and exposure to air it changes color progessively to gray, brown, and finally, almost black, occasionally with a pink tinge. Affected tissue has the smoky gray color of frosted tissue. After infection, a tuber may become so completely rotted (Plate 26) within a few days that even a slight pressure causes the skin to rupture and large quantities of liquid to exude. In the storage pile, all that remains of infected tubers are the tuber shells with thin papery skins (Fig. 43A). Cut seed tubers may also be rotted.

Causal Organisms

Pythium ultimum Trow., *P. debaryanum* Hesse, and possibly other *Pythium* spp. cause the disease. Oospores, which are smooth, thick-walled, and spherical, measure 14.2–19.5 μm and are terminal on branched coenocytic hyphae. Sporangia (Fig. 43B and C) are spherical, 12–29 μm, when produced terminally and barrel-shaped, 17–27 × 14–24 μm, when intercalary. Sporangia of *P. ultimum* do not produce zoospores. The mycelium is often difficult to isolate from diseased potato tissue.

Disease Cycle

The fungus lives in the soil and can enter tubers only through wounds. Infection, therefore, usually occurs at harvesting, grading, or less frequently at planting. Cut seed tubers are predisposed to infection after planting as soil temperatures begin to rise. Serious crop loss does take place in bruised, immature tubers harvested during hot, dry weather. The rot that develops is greatly aggravated by relatively high temperatures

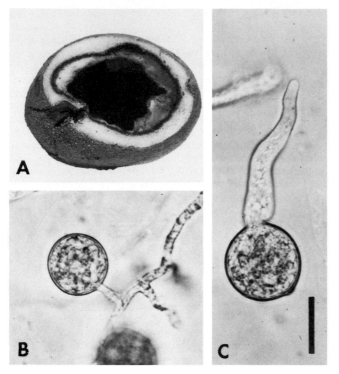

Fig. 43. Leak: **A,** typical tuber shell rot. *Pythium* sp. from rotted tuber: **B,** resting sporangium; **C,** spore with germ tube (bar represents 20 μm).

and poor ventilation but may be completely arrested under cool conditions.

Other Hosts

Both species of *Pythium* are pathogenic on an extremely wide range of hosts, including many market-garden crops, causing damping-off, root rot, or soft rot disease.

Control

1) Losses can be minimized by delay in lifting the crop so that tuber skins may mature.

2) Avoid mechanical injury to tubers by taking all possible care in harvesting.

3) If rotting begins in storage, increase air movement, and cool and dry the crop as quickly as possible.

Selected References

BUTLER, E. J., and S. G. JONES. 1949. Plant Pathology. Macmillan and Co., Ltd., London. 979 pp.

MIDDLETON, J. T. 1943. The taxonomy, host range and geographic distribution of the genus *Pythium*. Mem. Torrey Bot. Club 20:1–171.

TOMPKINS, C. M. 1975. World Literature on *Pythium* and *Rhizoctonia* Species and the Diseases They Cause. Contrib. No. 24, Reed Herbarium, Baltimore, MD.

WHITEHEAD, T., T. P. McINTOSH, and W. M. FINDLAY. 1953. The Potato in Health and Disease, 3rd ed. Oliver and Boyd, London. 744 pp.

(Prepared by A. R. McKenzie and C. H. Lawrence)

Pink Rot

Pink rot has been reported from nine states in the United States and from 11 other countries in North and South America, Europe the Middle and Far East, and Australia.

Symptoms

Wilting, which is sometimes the initial symptom, may occur at any time but generally occurs late in the season. Leaves become chlorotic, wilt, dry up, and abscise, starting at the stem base. Aerial tubers may form. Infection occurs in the roots, and all root, stolon, and stem tissues are killed as lesions develop. Stem lesions may extend up to the basal leaves, with water-soaking and light brown vascular discoloration at the advancing margin. Necrotic stems and roots are brown to black and may be confused with blackleg.

Tubers generally become infected (Plate 27) through diseased stolons, but some infections appear to occur at buds or lenticels. Rot proceeds uniformly through the tuber, with the advancing margin usually delimited by a dark line visible through the skin. Periderm over rotted portions is brownish-cream in white cultivars; tissues beneath lenticels are dark brown to black. Rotted tissues remain intact but are spongy. When recently or partially rotted tubers are cut, internal tissues are cream color, odorless, and rubbery or spongy in texture; if the cut tuber is squeezed, a clear liquid appears. On exposure to air, the color of infected tissues progressively changes to salmon pink (in 20–30 min) brown, and black (in about 1 hr). Internal tissues of tubers rotted for some time are black.

Causal Organism

Pink rot is most commonly caused by *Phytophthora erythroseptica* Pethybr. Sporangia are nonpapillate, highly variable in shape, ellipsoid or obpyriform, and 43 × 26 μm. Oogonia, 30–35 μm in diameter, have smooth walls 1 μm thick and may turn faintly yellow with age. Antheridia are amphigynous, ellipsoid or angular, and 14–16 × 13 μm. Oospores (Fig. 44) have walls 2.5 μm thick and nearly fill the oogonia.

The fungus grows on several media but not on dilute malachite green. Optimum and maximum growth temperatures are 24–28°C and 34°C, respectively. Asexual structures form in mycelial mats when transferred successively to mineral solution and water or when grown in water culture with boiled hemp seeds. Sexual organs form abundantly in agar media.

P. cryptogea Pethyb., *P. drechsleri* Tucker, *P. megasperma* Drechsler, and *P. parasitica* Dastur also infect potatoes and induce plant and/or tuber symptoms similar to those caused by *P. erythroseptica*. Single strains of *P. cryptogea* form sexual organs only sparingly after a few weeks in culture but promptly when grown with complementary strains of *P. cinnamomi*. The dimensions of reproductive structures of *P. cryptogea* are similar to those of *P. drechsleri,* but its optimal growth temperatures (22–25°C) are lower than those (28–31°C) for *P. drechsleri. P. megasperma* has paragynous antheridia, in contrast to amphigynous antheridia in the other species, and it has larger oospores (an average of 41 μm vs. about 24–25 μm). Sporangia of *P. parasitica* are papillate, and those of the other species are nonpapillate.

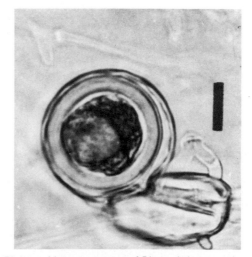

Fig. 44. Pink rot. Mature oospore of *Phytophthora erythroseptica* (bar represents 10 μm). (Courtesy R. C. Rowe and W. J. Hooker)

Disease Cycle

P. erythroseptica is soilborne and endemic in many soils. Zoospores, sporangia, or oospores may serve as inoculum, but oospores are probably the significant propagule in pathogen dissemination and survival in soil. Plants of all ages are susceptible, but the disease is most frequently observed in mature plants approaching harvest.

Epidemiology

Disease develops in soils approaching saturation from poor drainage or excessive precipitation or irrigation. Large amounts of decomposing plant residues in soil enhance water absorption, retention, and disease incidence. In wet soil, the disease develops over a wide range of temperatures but is most severe between 20 and 30° C.

Other Hosts

Although *P. erythroseptica* has been reported as a pathogen only on potatoes and tulips, it has been recovered from the roots of 17 nonsolanaceous plants, including wheat and rye.

Control

1) Plant seed tubers in soils with good drainage.
2) Avoid excessive irrigation late in the growing season.

Selected References

GOSS, R. W. 1949. Pink rot of potatoes caused by *Phytophthora erythroseptica* Pethyb. Neb. Agric. Exp. Stn. Res. Bull. 160. 27 pp.

ROWE, E. C., and A. F. SCHMITTHENNER. 1977. Potato pink rot in Ohio caused by *Phytophthora erythroseptica* and *P. cryptogea*. Plant Dis. Rep. 61:807–810.

STAMPS, D. J. 1978. *Phytophthora erythroseptica*. No. 593 in: Descriptions of Pathogenic Fungi and Bacteria. Commonw. Mycol. Inst., Kew, Surrey, England. 2 pp.

VARGAS, L. A., and L. W. NIELSEN. 1972. *Phytophthora erythroseptica* in Peru: Its identification and pathogenesis. Am. Potato J. 49:309–320.

WATERHOUSE, G. M. 1963. Key to the species of *Phytophthora* De Bary. No. 92 in: Mycol. Papers Commonw. Mycol. Inst., Kew, Surrey, England. 22 pp.

(Prepared by R. C. Rowe and L. W. Nielsen)

Late Blight

Late blight is probably the single most important disease of potatoes worldwide. It is destructive wherever potatoes are grown without fungicides, except in hot, dry, irrigated areas. The Irish "potato famine" of the 1840s was caused by *Phytophthora infestans*, the fungus that causes late blight. Immense quantities of fungicides are applied to potatoes throughout the world for protection against *P. infestans*.

Symptoms

Leaf lesions are highly variable, depending on temperature, moisture, light intensity, and host cultivar. Initial symptoms are typically small, pale to dark green, irregularly shaped spots. Under favorable environmental conditions they rapidly grow to large, brown to purplish black, necrotic lesions that may kill entire leaflets and spread via petioles to the stem, eventually killing the entire plant. A pale green to yellow halo is often present outside the area of leaf necrosis (Plate 28). Under moist conditions, a white downy mildew of sporangia and sporangiophores appears at the edge of lesions, mostly on the underside of the leaves (Plate 29).

In the field, plants severely affected with late blight give off a distinctive odor. This odor actually results from rapid breakdown of potato leaf tissue and also follows chemical vine killing, frosts, etc.

Positive identification of late blight requires confirmation of sporangia and sporangiophores either on lesions in the field under moist conditions or on leaf or tuber lesions incubated in a moist chamber.

On susceptible cultivars, exteriors of infected tubers show irregular, small to large, slightly depressed areas of brown to purplish skin (Plate 30). A tan-brown, dry granular rot characteristically extends into the tuber approximately 1.5 cm, the depth varying according to length of time after infection, cultivar, and temperature. The boundary between diseased and healthy tissue is not clearly defined; delicate, brown, peglike extensions penetrate to variable depths. Under cool, dry storage conditions, tuber lesions develop slowly and may become slightly sunken after several months. Secondary organisms (bacteria and fungi) often follow infection by *P. infestans*, resulting in partial or complete breakdown of tubers and complicating diagnosis.

Causal Organism

Phytophthora infestans (Mont.) de Bary has sporangia (conidia) that are hyaline, lemon-shaped, thin-walled, and 21–38 × 12–23 μm in size. Each has an apical papillum (Fig. 45C). Sporangia of *P. infestans* are borne on the tip of a sporangiophore branch (Fig. 45A); as it elongates, the sporangiophore swells slightly and turns the attached sporangium to the side. The sporangiopore is thus characterized by periodic swellings (Fig. 45B) at points where sporangia were produced.

Sporangia may germinate by means of a germ tube (Fig. 45E), but most commonly they form about eight biciliate zoospores (Fig. 45D) that swim freely in water and encyst on solid surfaces.

Encysted zoospores can germinate by germ tubes that enter the host via leaf stomata, but usually an appressorium is formed and penetration hyphae enter directly through the cuticle. Once inside the plant, the nonseptate mycelium is intercellular and intracellular by means of haustoria (Fig. 45G) that extend into cells.

Sexual reproduction results in oospores (Fig. 45F) formed by the union of oogonia and antheridia. Oospores within oogonia are 24–46 μm in diameter and germinate via a germ tube with a terminal sporangium, which, in turn, either liberates zoospores or forms another germ tube.

Disease Cycle

Oospores in nature have been found only in Mexico, where both mating types (A$_1$ and A$_2$) occur. Leaves touching the soil are often infected first, suggesting that oospores probably play a role in the survival of *P. infestans* under adverse conditions.

In tropical areas where the crop is grown all year, overwintering of *P. infestans* is not an important consideration. However, where distinct seasons occur, *P. infestans* overwinters as mycelium in unharvested tubers, tubers dumped in cull piles on farms or near commercial storages, or tubers stored and saved for seed. After plant emergence, the fungus invades a few of the growing sprouts and sporulates under moist conditions, producing primary inoculum. Once primary infection has occured, further spread of *P. infestans* takes place by airborne or waterborne sporangia (Fig. 46).

Tubers on cull piles frequently sprout and form dense masses of succulent tissue that are easily infected by *P. infestans* spores from diseased tubers. Sporulation within the foliage mass produces prodigious numbers of spores to infect nearby fields.

Epidemiology

Tubers, particularly those inadequately covered by soil, may be infected in the field by spores that have been washed from infected leaves into the soil by rain or irrigation. Rapid tuber growth frequently causes soil to crack, exposing tubers to infection. Tuber infection may also occur during wet harvest conditions via contact between tubers and sporangia on vines or via airborne sporangia. Little, if any, spread of *P. infestans* occurs under optimum conditions in storage.

Field infection is most successful under cool, moist conditions. However, infections take place over a range of environmental conditions, and high temperature strains of the

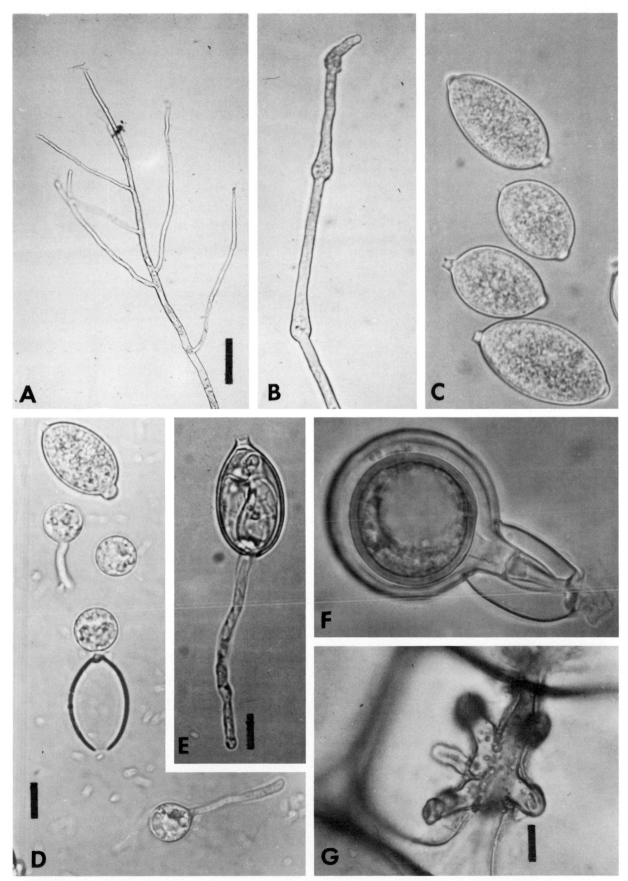

Fig. 45. Late blight. *Phytophthora infestans:* **A,** sporangiophore; **B,** sporangiophore branch showing swellings at successive sites of sporangium formation; **C,** sporangia, germinating by zoospores (**D**) and germ tube (**E**); **F,** oospore with antheridium; **G,** haustoria within tuber cells. Bar in A represents 50 μm; bars in D, E, and G represent 10 μm. (F, Courtesy of Plant Pathology Section, West Virginia University)

fungus have been reported. Sporangial production is most rapid and prolific at 100% rh and at 21° C. Sporangia are sensitive to desiccation and, after dispersal by wind or splashing water, require free water for germination. The optimal temperature for indirect germination via zoospores is 12° C, whereas that for direct germination of sporangia via germ tubes is 24° C. Both types of germination occur at overlapping temperatures, however. Zoospores, although quickly killed by drying, produce germ tubes and appressoria in the presence of free water. Penetration occurs at temperatures between 10 and 29° C. Once penetration has occurred, infection and subsequent development of disease is most rapid at 21° C.

Systems for forcasting late blight and for timing fungicide applications rely on records of temperature and rainfall (Hyre) or temperature and relative humidity (Wallin) and predict the probability of late blight development, assuming the presence of inoculum. A forecasting system combining both these systems is "Blitecast" (Krause et al), which is used in the northeastern United States for timing fungicide applications. Where rainfall and relative humidity are closely related, fungicides are applied after rainfall accumulated to 1.27 cm has theoretically washed previously applied fungicide from the foliage (Barriga et al).

Other Hosts

Late blight often severely affects tomatoes and occasionally affects eggplant and many other members of the Solanaceae.

Resistance

Two types of resistance to *P. infestans* in potatoes are recognized: 1) specific resistance (also called race specific, vertical, oligogenic, or monogenic resistance) and 2) general resistance (also called field, race nonspecific, horizontal, or polygenic resistance). Before the discovery of specific resistance, fairly high levels of general resistance were obtained. For several decades after discovery of specific resistance in *Solanum demissum,* breeders incorporated one or a few *S. demissum* genes into each new variety. Because *P. infestans* is highly variable, the pathogen rapidly overcame such resistance; use of specific resistance has therefore contributed little to controlling late blight. All potato cultivars and all tuber-bearing *Solanum* species are susceptible to late blight in the Toluca Valley of Mexico, where the sexual stage of *P. infestans* occurs; thus the probability of obtaining lasting specific resistance is very low. No cultivars in Europe or North American allow commercial cultivation of potatoes without fungicide protection. Some commercial cultivars, such as Sebago, have a moderate level of general resistance and are protected by lower amounts of fungicide than are required by other cultivars. Breeding efforts

on several continents are directed toward obtaining cultivars with high levels of generalized resistance that can be used with reduced amounts of fungicide or even without fungicide in drier areas.

Control

1) Avoid development of early season (primary) inoculum by the use of blight-free seed and destruction of potential inoculum sources such as cull piles, volunteer plants, etc.

2) Apply protectant fungicides as recommended by a forecasting service or (if such service is not available) as early as late blight is present in the area. Apply fungicides regularly as new vine growth develops and regularly after vines overgrowing the rows have caused high relative humidity within the canopy. Be sure that coverage of vines and leaves is thorough and uniform.

3) Prevent tuber infection by maintaining good soil coverage of tubers through adequate hilling. (Exceptionally large hills are commonly made in the Andes, resulting in relatively rare tuber infection.) Maintain adequate foliage protection to reduce inoculum production on leaves. Kill vines two weeks before harvest so that sporangia on leaves dry out and die and infected tubers rot, thus permitting identification and removal before the crop is placed in storage.

4) Prevent rot in storage by removing infected tubers before storage and maintaining adequate air circulation and temperature as cool as is compatible with other considerations.

5) Use resistant cultivars where possible.

Selected References

BARRIGA, R., H. D. THURSTON, and L. E. HEIDRICK. 1961. Ciclos de aspersion para el control de la "gota" de la papa. Agric. Trop. 17:616–622.

CROSIER, W. 1934. Studies in the biology of *Phytophthora infestans* (Mont.) de Bary. N. Y. Agric. Exp. Stn., Cornell, Mem. 155. 40 pp.

GALLEGLY, M. E. 1968. Genetics of pathogenicity of *Phytophthora infestans.* Ann. Rev. Phytopathol. 6:375–396.

GALLEGLY, M. E., and J. S. NIEDERHAUSER. 1959. Genetic controls of host-parasite interactions in the Phytophthora late blight disease. Pages 168–182 in: C. S. Holton, G. W. Fischer, R. W. Fulton, H. Hart, and S. E. A. McCallan, eds. Plant Pathology Problems and Progress, 1980–1958. Univ. Wis. Press, Madison. 588 pp.

HYRE, R. A. 1954. Progress in forecasting late blight of tomato and potato. Plant Dis. Rep. 38:245–253.

KRAUSE, R. A., L. B. MASSIE, and R. A. HYRE. 1975. Blitecast: A computerized forecast of potato late blight. Plant Dis. Rep. 59:95–98.

LARGE, E. C. 1962. The Advance of the Fungi. Dover Publ., New York. 488 pp.

WALLIN, J. R. 1962. Summary of recent progess in predicting late blight epidemics in the United States and Canada. Am. Potato J. 39:306–312.

(Prepared by H. D. Thurston and O. Schultz)

Powdery Mildew

Powdery mildew can be an important foliage disease in arid or semiarid climates. It has been reported from Chile, Peru, Mexico, New Zealand, Europe, and the Middle East. In the United States it is of economic importance only in the state of Washington under row irrigation, although it has been reported in Ohio and Utah.

Symptoms

Elongated, light brown stipples, 0.5–2 mm in length, may appear on stems and petioles of infected plants. These often coalesce to form larger, water-soaked, blackened areas on the petioles. Infections are initially powdery white (Fig. 47A, Plate 31) and later tan. Sporulation on both leaf surfaces appears as dusty, grayish-brown deposits that superficially resemble soil or spray residue. Severe infections may superficially resemble late

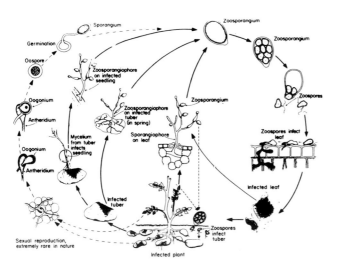

Fig. 46. Disease cycle of late blight caused by *Phytophthora infestans.* (Reprinted, by special permission, from Plant Pathology, 2nd ed., by G. N. Agrios. ©1978 Academic Press, New York)

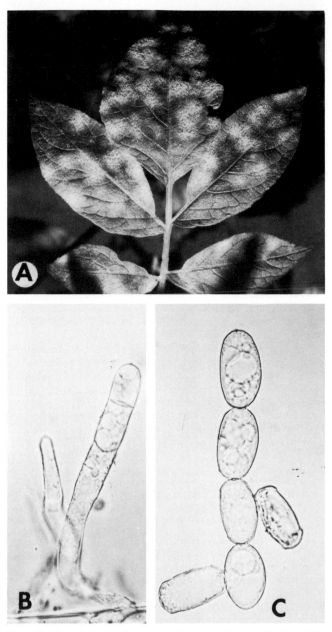

Fig. 47. Powdery mildew: **A**, early symptoms; **B**, *Erysiphe cichoracearum* conidiophore; **C**, mature conidia.

blight, with leaves becoming almost black (Plate 32), necrotic and abscising, leaving a rosette of upper foliage. Eventually, general infection can occur and the entire plant may collapse and die.

Causal Organism

Conidia of *Erysiphe cichoracearum* DC. ex Merat form in chains on unbranched conidiophores, $7-13 \times 36-50$ μm (Fig. 47B). Mature conidia (Fig. 47C) are oval to ellipsoid with flattened ends, $13-16 \times 20-30$ μm, and lack well-developed fibrosin bodies when mounted in water or 10% KOH. Cleistothecia have simple appendages and contain several asci, each usually containing two ascospores. Cleistothecia 135–165 μm in diameter with indeterminate appendages and 5–10 asci are very rare, having been reported only on field-grown potatoes from the western United States.

Differences in resistance exist among tuber-bearing *Solanum* spp. and within *S. tuberosum*.

Control

1) Dust or spray foliage with elemental sulfur at intervals of one to two weeks.

2) Powdery mildew is rarely a problem on potatoes grown under sprinkler irrigation. A heavy rain will also stop progress of the disease.

Selected References

DUTT, B. L., R. P. RAI, and H. KISHORE. 1973. Evaluation of reaction of potato to powdery-mildew. Indian J. Agric. Sci. 43:1063–1066.
ROWE, R. C. 1975. Powdery mildew of potatoes in Ohio. Plant Dis. Rep. 59:330–331.

(Prepared by R. C. Rowe and G. D. Easton)

Early Blight

This disease is found worldwide wherever potatoes are grown.

Symptoms

Initial infection is most frequent on lower, older leaves. Lesions first appear as small (1–2 mm) spots, dry and papery in texture, later becoming brown-black and circular-ovoid as they expand. Advanced lesions often have angular margins because of limitation by leaf veins. Concentric rings of raised and depressed necrotic tissue usually, but not always, give lesions a characteristic "target board" or "bullseye" appearance (Fig. 48A). Leaf tissue often becomes chlorotic around and among lesions. As new lesions develop and older ones expand, the entire leaf becomes chlorotic, later necrotic, and desiccates but usually does not abscise (Plate 33). Damage to leaves is considerably in excess of tissue actually destroyed by lesions, suggesting that toxins cause leaf death some distance from the site of infection. Advanced vine symptoms intergrade with those of Verticillium wilt and leaf scald associated with moisture stress in irrigated potatoes.

Tuber lesions are dark, sunken, circular to irregular in shape, and often surrounded by a raised border of purplish to gun metal color (Fig. 48B). The underlying flesh is dry, leathery to corky, and usually brown. Tissue in advanced decay is often water-soaked and yellow to greenish yellow. Lesions can increase in size during storage, and tubers can become shriveled in advanced cases. Early blight tuber lesions are not as prone to invasion by secondary organisms as are many other tuber rots.

Causal Organism

Alternaria solani Sorauer (syn. *Macrosporium solani* Ellis & Martin) has conidia $15-19 \times 150-300$ μm with 9–11 transverse septa and few, if any, longitudinal septa. Spores are usually borne singly but may be catenulate. They are straight or slightly bent, the body being ellipsoid to oblong and tapering gradually to a long beak (Fig. 48C and D). Color varies from pale to light tan to olive-brown. The beak is flexuous, pale, occasionally branched, and 2.5–5.0 μm wide. Conidiophores occur singly or in small groups and are straight or flexuous, pale to olive-brown, 6–10 μm in diameter and up to 100 μm long.

Cultural characteristics vary widely. Most isolates grow well on artificial media; however, they sporulate sparingly unless the mycelium is wounded or irradiated or they are cultured on a low nutrient medium. Colonies are spreading, hairy, and gray-brown to black. Some isolates produce a yellowish red pigment in nutrient media.

Disease Cycle

Depending upon the location, *A. solani* persists in crop debris, soil, infected tubers, or other solanaceous hosts. The fungus penetrates the leaves directly through the epidermis. Primary infection can occur on older foliage early in the season. However, actively growing young tissue and plants heavily fertilized with nitrogen do not exhibit symptoms, and most secondary spread occurs as plants age, especially after blossoming, when secondary inoculum levels are higher. In many locations, early blight is principally a disease of senescing plants.

Immature tuber surfaces are easily infected, whereas those of mature tubers are much more resistant. Wounds are generally necessary for infection through mature tuber skins. A period of 3–4 days or more between vine killing and digging considerably increases tuber resistance.

Epidemiology

Maximum mycelial growth of *A. solani* in pure culture occurs at 28° C, whereas optimum temperature for formation of conidiophores and conidia is 19–23° C. Conidiophore formation is inhibited, but not irreversibly, at temperatures greater than 32° C. Temperatures above 27° C stop conidia formation. Conidiophores develop in light, whereas light inhibits conidia formation at temperatures above 15° C. Maximum spore production in the field occurs between 3:00

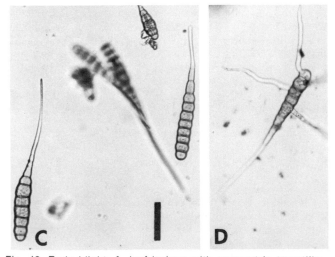

Fig. 48. Early blight: **A,** leaf lesions with concentric targetlike markings, somewhat limited by the larger veins; **B,** tuber lesions; **C,** dormant spores of *Alternaria solani* with short-beaked *A. alternata* type near top (bar represents 50 μm); **D,** germinating spore.

a.m. and 9:00 p.m. Spores in water germinate within 35–45 min at the optimum temperature (24–30° C) and within 1–2 hr at 6–34° C. Optimum temperature for tuber infection is 12–16° C but varies with cultivar.

Most rapid progress of the disease occurs during periods of alternating wet and dry weather. Early blight can be severe in irrigated desert regions because of prolonged periods of dew. The disease is often more severe when the host has been predisposed by injury, poor nutrition, or other type of stress.

Field resistance to foliage infection is associated with plant maturity. Late maturing varieties are usually more resistant. Early blight does not reduce yields when infection occurs late in the season.

Other Hosts

The fungus is pathogenic on tomato and other solanaceous crops and has been reported on other genera such as *Brassica* spp.

Control

1) Cultivars with levels of field resistance are available, but no cutivars are immune.

2) Protectant fungicides such as the dithiocarbamates, fentin hydroxide, and chlorothalonil effectively control early blight on foliage.

3) Fungicide applications scheduled by spore trapping or other methods so as to coincide with secondary spread of the disease are most effective. Early season applications of fungicides before secondary inoculum is produced often have little or no effect on the spread of the disease.

4) Permit tubers to mature in the ground before digging, and avoid bruising in handling.

5) Avoid disturbing seed tubers until ready to plant.

Selected References

DOUGLAS, D. R., and M. D. GROSKOPP. 1974. Control of early blight in eastern and southcentral Idaho. Am. Potato J. 51:361–368.
DOUGLAS, D. R., and J. J. PAVEK. 1972. Screening potatoes for field resistance to early blight. Am. Potato J. 49:1–6.
ELLIS, M. B., and I. A. S. GIBSON. 1975. *Alternaria solani.* No. 475 in: Descriptions of Pathogenic Fungi and Bacteria. Commonw. Mycol. Inst., Kew, Surrey, England. 2 pp.
HARRISON, M. D., C. H. LIVINGSTON, and N. OSHIMA. 1965. Control of potato early blight in Colorado. I. Fungicidal spray schedules in relation to the epidemiology of the disease. Am. Potato J. 42:319–327.
HARRISON, M. D., C. H. LIVINGSTON, and N. OSHIMA. 1965. Control of potato early blight in Colorado. II. Spore traps as a guide for initiating applications of fungicides. Am. Potato J. 42:333–340.
VENETTE, J. R., and M. D. HARRISON. 1973. Factors affecting infection of potato tubers by *Alternaria solani* in Colorado. Am. Potato J. 50:283–292.
WAGGONER, P. E., and J. G. HORSFALL. 1969. Epidem. A simulator of plant disease written for a computer. Conn. Agric. Exp. Stn. Bull. 698. 80 pp.

(Prepared by D. P. Weingartner)

Alternaria alternata

Alternaria alternata (Fries.) Keissler (syn. *A. tenuis* Nees.) infects potato and other solanaceous crops, forming lesions on potato leaves similar to those of early blight. Spores (20–63 × 9–18 μm) are smaller than those of *A. solani,* are formed in chains, and lack the typical long beak (Fig. 49A). Their size and shape may vary considerably. The fungus is often associated with other diseases and is frequently isolated. It is generally considered a weak parasite, which attacks plants weakened by viruses, deficiencies, stress, or senescence.

Selected References

ELLIS, M. B. 1971. Dematiaceous Hyphomycetes. Commonw. Mycol.

Inst., Kew, Surrey, England. 608 pp.

SALZMANN, R., and E. R. KELLER. 1969. Krankheiten und Schädlinge der Kartoffel. Landwirtschaftliche Lehrmittelzentrale Zollikofen. 150 pp.

SREEKANTIAH, K. R., K. S. NAGARAJA RAO, and T. N.

RAMACHANDRA RAO. 1973. A virulent strain of *Alternaria alternata* causing leaf and fruit spot of chilli. Indian Phytopathol. 26:600–603.

(Prepared by L. J. Turkensteen)

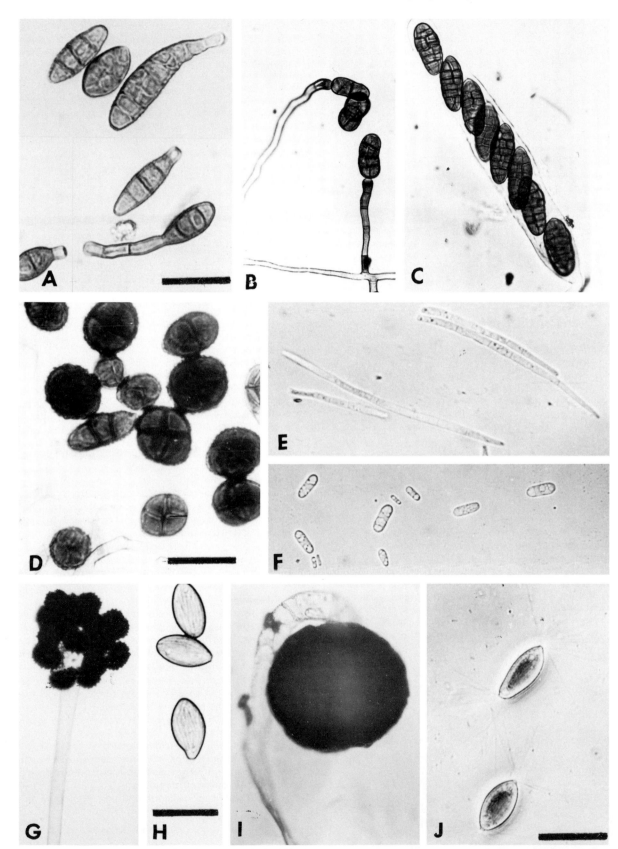

Fig. 49. Representative structures for identification of: **A,** *Alternaria alternata,* conidia; **B,** *Pleospora herbarum (Stemphylium botryosum),* conidia and **(C)** ascospores in ascus; **D,** *Ulocladium atrum (Stemphylium atrum),* conidia; **E,** *Septoria lycopersici,* conidia; **F,** *Phoma andina,* spores from plate culture; **G–J,** *Choanephora cucurbitarum,* sporangiophores and conidia of two distinct types. Bars represent 20 μm. (B and C, Courtesy Research Institute for Plant Protection, Wageningen; A and D–J, courtesy L. J. Turkensteen and W. J. Hooker)

Pleospora herbarum

Pleospora herbarum (Pers. ex Fr.) Rabenh. and its imperfect stage, *Stemphylium botryosum* Wallr., are often found associated with potato plants poorly adapted to warm conditions or to other environmental stresses. Round, rapidly enlarging, light colored leaf lesions develop; affected tissues appear as thin paper (Plate 34). The same fungus is commonly isolated from dead and dried materials and wood, but is also known as a pathogen for several crops.

Conidia are olive to brown and oblong, with three transverse septa and 1–3 longitudinal septa. The spores are $19.5 \times 28.5 \, \mu m$, with a single basal pore $8 \, \mu m$ in diameter (Fig. 49B). Ascospores are dark, yellow-brown, ellipsoid to clavate, muriform, and $26–50 \times 10–20 \, \mu m$ (Fig. 49C). Transmission is by airborne ascospores or conidia, and penetration is through stomata.

Selected References

BOOTH, C., and K. A. PIROZYNSKI. 1967. *Pleospora herbarum*. No. 150 in: Descriptions of Pathogenic Fungi and Bacteria. Commonw. Mycol. Inst., Kew, Surrey, England. 2 pp.

ELLIS, M. B. 1971. Dematiaceous Hyphomycetes. Commonw. Mycol. Inst., Kew, Surrey, England. 608 pp.

(Prepared by L. J. Turkensteen)

Ulocladium Blight

Ulocladium atrum Preuss (syn. *Stemphylium atrum* (Preuss) Sacc.) is a weakly pathogenic organism (Fig. 49D). In the high (over 3,500 m) Andean region around Lake Titicaca, it causes a potato foliage blight that is associated with damage from insects and, especially, from hail. Damage caused by the frequent hail storms is increased considerably as torn edges of leaves turn dark to black. When heavily attacked by *U. atrum,* the whole foliage turns blackish and becomes necrotic. Also, tiny, dark colored lesions up to 3 mm in diameter with irregular margins are formed on healthy leaves, apparently without previous wounding. Peruvian natives call the disease kasahui.

Selected References

FRENCH, E. R., H. TORRES, T. A. de ICOCHEA, L. SALAZAR, C. FRIBOURG, E. N. FERNANDEZ, A. MARTIN, J. FRANCO, M. M. de SCURRAH, I. A. HERRERA, C. VISE, L. LAZO, and O. A. HIDALGO. 1972. Enfermedades de la papa in el Perú. Bol. Tecn. No. 77. Est. Exp. Agríc. La Molina. 36 pp.

ELLIS, M. B. 1976. More Dematiaceous Hyphomycetes. Commonw. Mycol. Inst., Kew, Surrey, England. 507 pp.

(Prepared by L. J. Turkensteen)

Fig. 50. Leaf spot lesions of *Septoria lycopersici*. (Courtesy E. R. French)

Stemphylium consortiale

Stemphylium consortiale (Thüm.) Groves & Skolko (syn. *Ulocladium consortiale* (Thüm.) Simmons) causes lesions occurring with and superficially resembling those caused by *Alternaria solani*, except that lesions lack the concentric markings of early blight and are lighter brown. Lesions develop three to four days after inoculation and cause defoliation similar to that of early blight.

Selected References

ELLIS, M. B. 1976. More Dematiaceous Hyphomycetes. Commonw. Mycol. Inst., Kew, Surrey, England. 507 pp.

WRIGHT, N. S. 1947. A stemphylium leaf spot on potatoes in British Columbia. Sci. Agric. 27:130–135.

(Prepared by W. J. Hooker)

Septoria Leaf Spot

The disease is present in Central and South America. It occurs in cultivated potatoes at elevations that differ considerably from one region to another, e.g., at 1,600–2,500 m in Venezuela and at 3,800–4,200 m in Peru. It occurs in wild potato at a wider range of elevations.

Symptoms

Lesions on leaves are round to oval (Fig. 50) and have concentric rings of raised tissue when viewed from the upper surface. Rings are similar to those of early blight or Phoma leaf spot. Septoria leaf spot can be distinguished with the help of a good hand magnifier by the presence of one or more relatively large, erumpent pycnidia ($90–230 \, \mu m$) in older lesions. Leaves in a late stage of attack become brittle, deformed, and susceptible to wind damage. In advanced stages, leaves become necrotic and may drop from the plant. Yield reductions are considerable.

Causal Organism

Septoria lycopersici Speg. is similar on tomato and potato plants but exhibits differences when isolates from the two hosts are grown on artificial media. Dampened pycnidia release masses of hairlike spores ($1.8–2.4 \times 25–135 \, \mu m$ or longer) with three or four, sometimes up to seven, cross walls (Fig. 49E). On artificial media, lead gray colonies expand very slowly, and oatmeal agar becomes brown below the colony.

Epidemiology

The disease is present in regions characterized by cool, moist weather during the growing season. Inoculum is transported by rain splash and probably carries over on plant debris in soil. Long moist periods during which leaves stay wet are thought to be necessary for infection.

Other Hosts

Tomato is the principal other host.

Control

1) Nonsystemic fungicides capable of controlling late blight are effective agaist *Septoria*. Treatment should be started at an early stage of infection because lesions, once present, form a continuous source of inoculum.

2) Differences in susceptibility have been observed.

Selected References

PIGLIONICA, V., G. MALAGUTI, A. CICCARONE, and G. H. BOEREMA. 1979. La Septoriosi della patata. Phytopathol. Mediterr. 17:81–89.

JIMENEZ, A. T., and E. R. FRENCH. 1972. Mancha anular foliar (*Septoria lycopersici* subgrupo A) de la papa. Fitopatologia 5:15–20.

TORRES, H., E. R. FRENCH, and L. W. NIELSEN. 1970. Potato diseases in Peru, 1965–1968. Plant Dis. Rep. 54:315–318.

(Prepared by L. J. Turkensteen)

Cercospora Leaf Blotches

The disease is reported from cool and temperate climates of Europe and Russia and from the eastern part of the United States, where it is not considered an important disease. It is also reported from restricted areas in Africa and Asia and from India, where it occurs with early and late blights.

Symptoms

First symptoms on lower leaves are small yellow to purplish lesions that increase from 0.2 to 1 cm in size. On the underside of the lesions, a dense, plush, gray layer of conidiophores and conidia is formed. Later, lesions are separated from surrounding tissues by a dark line. When lesions become necrotic, tissue may drop out, leaving only holes. Necrotic lesions are distinguished from those of *Alternaria solani* by the lack of concentric rings. The disease becomes apparent at about the same time as late blight. The leaf may be killed; stem lesions become dark; and the entire plant may die. Symptoms on tubers have not been described.

Causal Organism

Mycovellosiella (Cercospora) concors (Casp.) Deighton has dark spores formed on densely branched sporophores that emerge through stomata. The straight or slightly bent, dark spores are variable in length ($14-57 \times 3.5-6$ μm) and may have up to six septa or none (Fig. 51).

An additional *Cercospora* species can attack potato. This larger spored form, *C. solani-tuberosi* Thirumalachar, with conidia $41-120 \times 3.3$ μm (1–12 septate), is described from India.

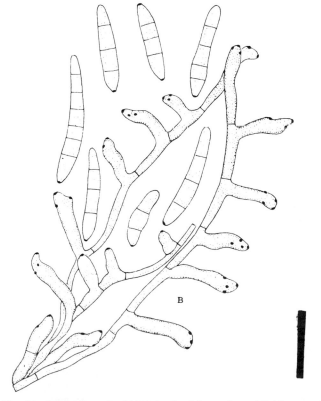

B

Fig. 51. Cercospora leaf blotch. Conidia and conidiophores of *Mycovellosiella (Cercospora) concors*. Bar represents 25 μm. (Reprinted, by special permission, from More Dematiaceous Hyphomycetes, by M. B. Ellis. ©1976 Commonwealth Mycological Institute, Kew, Surrey, England)

Control

Foliage sprays used for other leaf pathogens are apparently effective.

Selected References

ELLIS, M. B. 1976. More Dematiaceous Hyphomycetes. Commonw. Mycol. Inst., Kew, Surrey, England. 507 pp.
NAGAICH, B. B., G. S. SHEKHAWAT, S. M. KHURANA, and S. K. BHATTACHARYA. 1974. Pathological problems of the potato cultivation in India. J. India Potato Assoc. 1:32–44.
SALZMANN, R., and E. R. KELLER. 1969. Krankheiten und Schädlinge der Kartoffel. Landwirtschaftliche Lehrmittelzentrale Zollikofen. 150 pp.
THIRUMALACHAR, M. J. 1953. Cercospora leaf spot and stem canker disease of potato. Am. Potato J. 30:94–97.

(Prepared by L. J. Turkensteen)

Phoma Leaf Spot

Yield losses up to 80% have occurred from Phoma leaf spot in the Andes of Peru and Boliva at altitudes of 2,000–3,500 m.

Symptoms

Small leaf spots, up to 1 cm but mostly less than 2.5 mm in diameter, have concentric rings and are similar to early blight lesions except that the lesions are not depressed into the leaf tissue (Plate 35). At first a few lesions form on the lower leaves; gradually infection spreads to the whole plant. Primary lesions continue to expand, and secondary infections give rise to many smaller lesions, which may coalesce. Foliage becomes blackish and when it dies remains attached to the stem for a time before it drops. On stems and petioles, elongate lesions develop.

Causal Organism

Phoma andina Turkensteen has light-colored pycnidia, 125–200 μm in diameter (on artificial media) containing a distinct ostiole surrounded by 2–3 rows of brown cells. In leaf lesions, many submerged pycnidia are present at the upper side only. From the same pycnidium, two types of spores are formed: 1) hyaline, one-celled infective spores, $14-22 \times 5-7$ μm, shaped like two-seeded peanut pods (Fig. 49F), and 2) small sporelike bodies, $5.8-7.8 \times 2.0-2.6$ μm, which do not germinate on artificial media nor infect plants. On artificial media, single, hyaline chlamydospores develop in series, but complex chlamydospores are also formed. In rare cases, spores in old cultures may be two-celled and considerably larger.

The colony is light colored on artificial media. It is relatively slow growing, and on acid (pH 4.5) media its growth is strongly inhibited. On potato-dextrose agar and oatmeal agar the medium turns yellow-green and yellow, respectively, within two to three weeks. When grown on slightly acid agar for one week, the medium turns yellow when a drop of $1N$ NaOH is added to the surface.

Other Hosts

Cultivated and wild species of potato are as yet the only known hosts.

Control

1) Applications of fungicides are effective when started early in the season before lesions are abundant.
2) Resistance is known.

Selected References

TORRES, H., E. R. FRENCH, and L. W. NEILSEN. 1970. Potato diseases in Peru, 1965–1968. Plant Dis. Rep. 54:315–318.

TURKENSTEEN, L. J. 1978. Tizon foliar de la papa en el Perú: I. Especies de Phoma asociadas. Fitopatologia 13:67–69.

(Prepared by L. J. Turkensteen)

Choanephora Blight

Choanephora blight, caused by *Choanephora cucurbitarum* (Berk. & Rav.) Thaxter is only known so far in hot, moist, tropical sites in Peru where potato has recently been introduced. It is marked by long (4–5 mm), spiny sporangiophores (Fig. 49G–J) on initially water-soaked, but later necrotic, lesions (Fig. 52). Affected plants may die rapidly or slowly.

Selected Reference

TURKENSTEEN, L. J. 1979. Choanephora blight of potatoes and other crops grown under tropical conditions in Peru. Neth. J. Plant Pathol. 85:85–86.

(Prepared by L. J. Turkensteen)

Gray Mold

The fungus is found on a wide range of plants throughout the world but in potato the disease is usually considered of minor economic importance.

Symptoms

Symptoms become apparent on foliage toward the end of the growing season. Lesions on upper leaves are rare, developing only during periods of cool weather. Lesions are usually on the margins or tips of leaves, typically wedge-shaped, often bordered by major veins, and have wide concentric zonation. They may superficially resemble late blight lesions. When infected flower parts fall on leaves, the fungus may grow from the decaying part and produce a somewhat circular lesion (Fig. 53B).

Lower leaves that have become chlorotic from shading break down with a slimy rot. Rot spreads from infected leaves through the petiole and into the cortex of the stem (Fig. 53A).

Botrytis fruits profusely on affected tissue, producing a fuzzy appearance. Spore masses and aerial mycelia with gray mold are relatively dense and off-white or gray to tan in contrast to those with late blight, which are sparse and white.

Tuber infection, which is uncommon, is not apparent at digging but develops during storage and may become severe. The surface of infected tissue is wrinkled. Underlying tissue is flabby, temporarily darkened, and later becomes semiwatery with brown decay. Tufts of the fungus may emerge from wounds and eyes (Fig. 53C and D). A dry type of rot also develops,

Fig. 52. *Choanephora cucurbitarum* infection on potato leaves. (Courtesy L. J. Turkensteen)

appearing as sunken, pitted, discolored areas penetrating usually less than 1 cm.

Causal Organism

Botrytis cinerea Pers. produces conidia in a grapelike cluster. conidia are ellipsoid to ovoid, one-celled, 9–15×6.5–10 μm, and borne on the tips of condiophores (Fig. 53E–G). Botrytis fruits on decaying petioles, stems, flowers, and tuber parts, and on sclerotia. Sclerotia are hard, black, irregularly shaped, 1–15 mm long, and firmly attached to the substrate.

The perfect stage, *Sclerotinia fuckeliana* (de Bary) Fuckel (syn. *Botryotinia fuckeliana* (de Bary) Whetz.) is relatively rare, with apothecia 1.5–7.0 mm in diameter and 3–15 mm high.

Epidemiology

Infection, sometimes initially latent, becomes apparent on senescent plant parts under stress from shading or excessive humidity. Spores are spread by wind and rain, and lesion development is limited by dry sunny conditions. Leaf infection requires high humidity and relatively cool temperatures. Inoculum is apparently ubiquitous.

High levels of K and N fertility reduce the percentage of tuber infection. Relative maturity of tubers has little influence on incidence of infection. Tuber decay may be severe when tubers are placed immediately into storage at moderately low temperatures and high humidity without previous wound healing.

Control

1) Foliage protectant sprays may be useful if foliage is not severely shaded.

2) Permit tubers to wound heal before placing them in low temperature storage.

Selected References

ELLIS, M. B., and J. M. WALLER. 1974. *Sclerotinia fuckeliana* (conidial state: *Botrytis cinerea*). No. 431 in: Descriptions of Pathogenic Fungi and Bacteria. Commonw. Mycol. Inst., Kew, Surrey, England. 2 pp.

HARPER, P. C., and H. WILL. 1968. A response of grey mold of potatoes to fertilizer treatment. Eur. Potato J. 11:134–136.

HOLLOMON, D. W. 1967. Observations on the phylloplane flora of potatoes. Eur. Potato J. 10:53–61.

RAMSEY, G. B. 1941. *Botrytis* and *Sclerotinia* as potato tuber pathogens. Phytopathology 31:439–448.

(Prepared by W. J. Hooker)

White Mold

White mold is a cool temperature disease occurring in the Andes and in temperate zones nearly everywhere that potatoes are grown.

Symptoms

Water-soaked lesions covered by a cottony mycelial mat and sclerotia are most frequent on the main stem at the soil line or on lateral branches in contact with the soil. They may appear, however, in the angles of secondary branches, leaves, petioles, and flower peduncles. Leaf lesions, irregularly shaped, are usually at the base of leaflets. First symptoms are small areas of discolored tissue that turn gray and look wet. In severely affected plants, the stem is girdled and plants die (Plate 36). Infected stems often with zonate lesions (Fig. 54A) have mycelia and sclerotia in the pith (Fig. 54B).

Tubers near the soil surface may be infected, beginning with small depressed areas sometimes located near the eyes and having a distinct demarcation between affected and healthy tissues. As lesions enlarge, flesh shrinks and becomes superficially blackened and spongy (Plate 37). Rot is watery and soft, becoming leaky under pressure. It is whiter than normal and, even in advanced stages, discolors the tuber only slightly

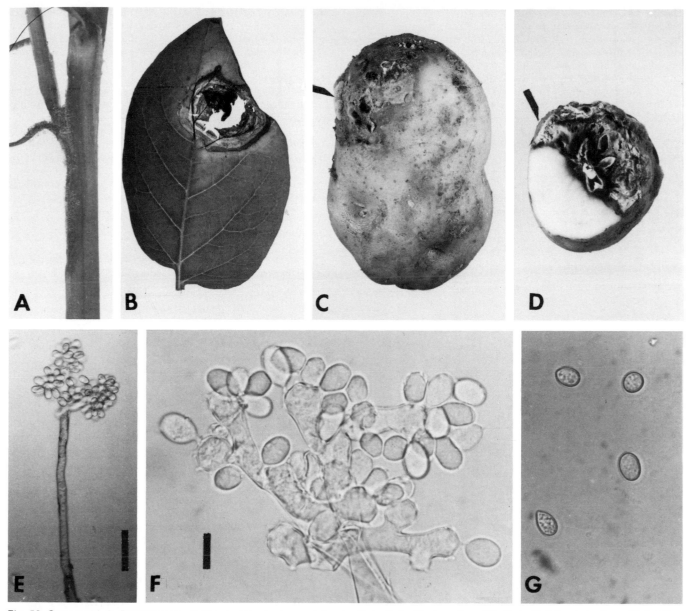

Fig. 53. Gray mold. **A,** *Botrytis* sporulating on stem lesion and dead petiole; **B,** leaf lesion; **C** and **D,** *Botrytis* tuber infection with feltlike mass of mycelium at arrow; **E** and **F,** *Botrytis cinerea* conidia on conidiophore; **G,** conidia. Bar in E represents 50 µm; bar in F represents 10 µm.

and gives no odor. Infected tissues later develop internal cavities filled with mycelium and sclerotia.

Causal Organisms

Sclerotinia sclerotiorum (Lib.) de Bary (syn. *Whetzelinia sclerotiorum* (Lib.) Korf. & Dumont) and also *S. minor* Jagger and *S. intermedia* Ramsey are causal agents. Sclerotia (Fig. 54C) are hard, lenticular or irregular in shape, and varied in size. They may be minute or may be over a centimeter in diameter. Young sclerotia have black coverings and white interiors; later the interior becomes black and firm. The mycelium is white and fluffy. Minute, globose spermatia (2–3 µm) develop infrequently in mycelial mats at the base of the plant or in drying cultures. Their role in pathogenicity is unknown. Apothecia (Fig. 55 top) formed at the soil surface may be funnel-shaped to flat. Their colors include pale orange, pink, light tan, and white. Apothecia can be 0.5 cm or more in diameter, and one or several may emerge from a single sclerotium. Asci are eight-spored, hyaline, cylindroclavate, 80–250 × 4–23 µm. Ascospores are unicellular, 6–28 × 2–15.2 µm, ovoid, hyaline, narrow at the base, and slightly inflated at the apex. Paraphyses are filiform and hyaline (Fig. 55 bottom).

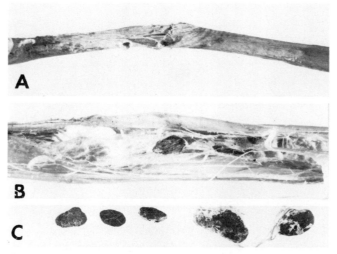

Fig. 54. White mold: **A,** zonate lesions on stem; **B,** stem interior showing cavity containing *Sclerotinia sclerotiorum* mycelium and a sclerotium; **C,** sclerotia.

49

Disease Cycle

Soilborne sclerotia near the surface germinate, forming either an apothecium or, if enough moisture and organic matter are available, a mycelial mat. The mycelium penetrates stems at the soil line and forms a white, fluffy mat on the stem and, frequently, also on adjacent soil. The fungus invades plant tissues rapidly, entering the inner stem tissues and pith, where sclerotia are formed. Apothecia forcefully eject numerous ascospores at maturity. Ascospores may spread a considerable distance from the source, settle on lateral branches or leaf surfaces, germinate, infect, and cause lesions. The fungus overwinters as sclerotia in the soil and in crop residues.

Epidemiology

Cool temperatures (16–22°C) and high relative humidity (95–100%) favor disease development. Sclerotia are killed within 3–6 weeks in flooded fields. Older tissue appears to be more susceptible than young tissue because the disease spreads more rapidly after plants are flowering and forming tubers. Heavy rainfall or irrigation induces apothecia production from sclerotia. Ejected ascospores are more effective in disease dissemination than is mycelium from sclerotia.

Other Hosts

S. sclerotiorum has a wide host range, attacking many dicotyledonous crops and weeds. Among solanaceous plants, potato, tomato, pepper, tobacco, and eggplant are severely attacked.

Control

1) Crop rotation with graminaceous crops for four or more years reduces disease incidence.

2) Fungicide application, especially with systemics, has been reported to give good control.

3) Flooding fields between crops may destroy sclerotia.

Selected References

BUSTAMENTE, E. R., and H. D. THURSTON. 1965. Pudrición dura del tubérculo de la papa. Agric. Trop. 21:113–121.

EDDINS, A. H. 1937. Sclerotinia rot of Irish potatoes. Phytopathology 27:100–103.

MOORE, W. D. 1949. Flooding as a means of destroying the sclerotia of *Sclerotinia sclerotiorum*. Phytopathology 39:920–927.

PARTYKA, E. E., and W. F. MAI. 1962. Effects of environment and some chemicals on *Sclerotinia sclerotiorum* in laboratory and potato field. Phytopathology 52:766–770.

PURDY, L. H. 1955. A broader concept of *Sclerotinia sclerotiorum* based on variability. Phytopathology 45:421–427.

RAMSEY, G. B. 1941. *Botrytis* and *Sclerotinia* as potato tuber pathogens. Phytopathology 31:439–448.

(Prepared by T. A. de Icochea)

Stem Rot

Stem rot affects potatoes in the tropical and subtropical regions. It is also reported from some countries with temperate climates: New Zealand, Denmark, The Netherlands, Argentina, Chile, the United States, and Russia.

Symptoms

Plant stems are infected at or below the soil surface (Plate 38). The plants wilt and lower leaves become chlorotic. An appressed, white, fanlike mycelial growth radiates over the soil surface, and numerous round, tan sclerotia form in the older mycelia at the stem base and soil surface. Lesions usually grow up and down the stem, and all living tissues are killed. Initially, infected tissues are soft, depressed, and brownish. As the dead cortical stem tissues dry out, xylem remains as fibrous strands.

Tubers become infected through the stolons of diseased plants and through lenticels from mycelia growing over tuber surfaces. The fungus radiates, forming symmetrical circles around the lenticels. Fresh lenticel lesions are moist, semifirm, and cheesy. They are easily dislodged and leave a cavity. After drying, the circular lesions become white and chalky (Fig. 56B). Multiple lesions may form on the tubers, destroying them before harvest. Secondary invaders, *Erwinia* spp., enter through these lesions and accelerate tuber decay. Infections incipient at harvest continue rotting during transit and storage, and often white superficial, closely appressed, mycelial strands radiate from the infection site (Fig. 56A).

Natural infection of seed tubers in the field occasionally causes seed decay and reduced stands.

Dead or dying plants devoid of fungus signs can often be diagnosed by placing them in a moist chamber for a few days; abundant mycelia grow from them.

Causal Organism

The mycelium of *Sclerotium rolfsii* Sacc. is white when young, becoming tan as it gets older. It is 6–9 μm in diameter and has thick clamp connections (Fig. 56C). Older mycelia usually form

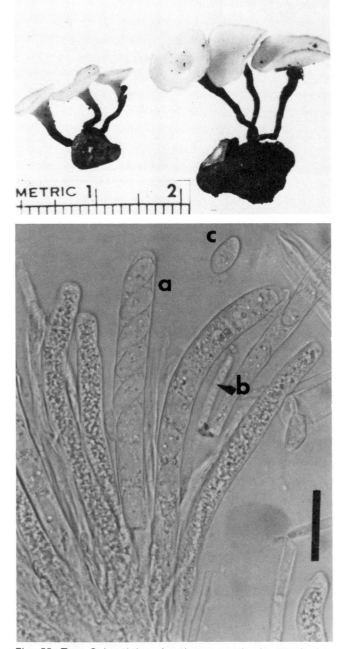

Fig. 55. **Top,** *Sclerotinia sclerotiorum* apothecia attached to sclerotia. **Bottom,** Hymenial layer of ascus: **a,** ascus with spores; **b,** paraphyses; **c,** an ascospore. Scale at top is in centimeters; bar represents 20 μm. (A, Courtesy T. A. de Icochea)

strands of pigmented hyphae. Sclerotia are numerous, round, 0.4–2.0 mm in diameter, white when young, then tan, and dark brown when old. The basidial stage, *Pellicularia rolfsii* (Sacc.) West, is uncommon, with basidiospores approximately 3.5–5 × 6–7 μm, elliptical to obovate, rounded above and either rounded or pointed at the base. Basidiospores do not seem to be important in the disease cycle.

Histopathology

Hyphae are both intracellular and intercellular and are constricted at the point of penetration of the cell wall. Host cells are killed well in advance of fungus hyphae. The hyphae produce oxalic acid in considerable quantities and also the enzymes polygalacturanase and cellulase, which hydrolyze and disrupt cell walls.

Disease Cycle

The fungus is soilborne as sclerotia or mycelia on decaying vegetable matter. Sclerotia permit long-term survival but contain relatively low energy reserves. They produce short-lived mycelia unless a suitable living or dead plant part is available. Mycelia infect seed tubers, sprouts, grown plants at any stage, and tubers. As an energy supply is exhausted, the mycelia aggregate, and sclerotia are formed. Disease spread in the field is by mycelial growth or by dispersal of mycelial fragments and sclerotia in debris or infested soil. Spread is therefore slow. Disease incidence within a field is often erratic, with infected plants as foci. Long distance spread is by transfer of infected plant parts containing hyphae or sclerotia, by movement of sclerotia by wind or surface water, or by mechanical means.

Epidemiology

Germination of sclerotia and mycelial growth are favored by aerobic conditions, high temperatures (28–30° C), and high relative humidity. Vegetative survival is in the upper few centimeters of the soil but may be deeper if the soil is dry and well aerated. Cool climates (elevations above 1,000 m in the tropics) are not favorable for disease development.

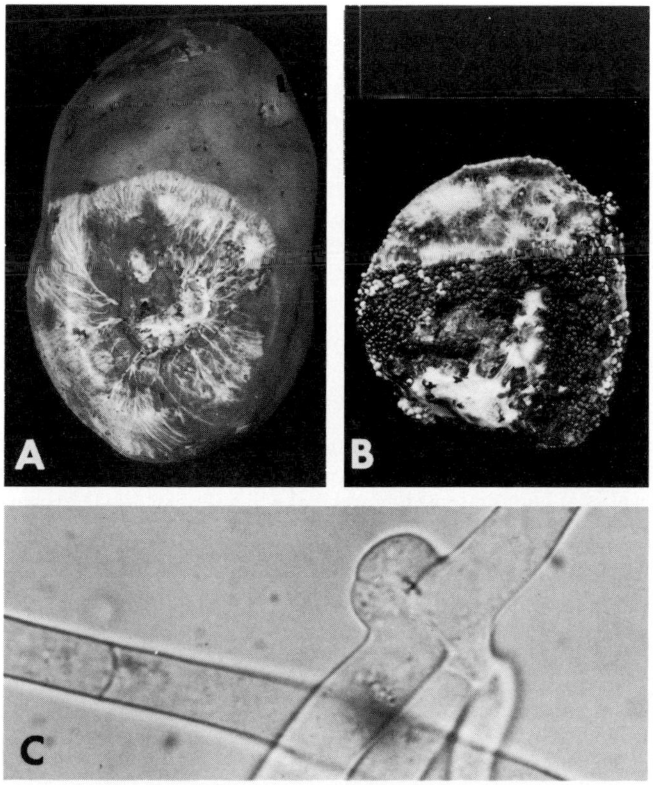

Other Hosts

S. rolfsii infects cultivated and noncultivated plants such as ferns, certain mosses, gymnosperms, grasses, cereals, banana, and many dicotyledonous plants including certain woody trees. It also grows on plant residues, including wood, in appropriate temperature and relative humidity.

Control

1) Use pentachloronitrobenzene as tuber seed treatment.
2) Fumigate soil with 31% sodium *N*-monomethyldithiocarbamate dihydrate.
3) Bury plant debris by deep plowing.
4) Avoid throwing soil or organic material (weeds) onto lower stems of potato plants during cultivation.
5) Control defoliating diseases to prevent accumulation of leaves on soil surfaces.

Selected References

AYCOCK, R. 1966. Stem rot and other diseases caused by *Sclerotium rolfsii*. N. C. Agric. Exp. Stn. Tech. Bull. 174. 202 pp.
AYCOCK, R., chairman. 1961. Symposium on *Sclerotium rolfsii*. Phytopathology 51:107–128.
EDDINS, A. H., and E. WEST. 1946. Sclerotium rot of potato seed pieces. Phytopathology 36:239–240.
FRENCH, E. R., H. TORRES, T. A. de ICOCHEA, L. SALAZAR, C. FRIBOURG, E. N. FERNANDEZ, A. MARTIN, J. FRANCO, M. M. de SCURRAH, I. A. HERRERA, C. VISE, L. LAZO, and O. A. HIDALGO. 1972. Enfermedades de la Papa en el Perú. Bol. Tecn. No. 77. Est. Exp. Agric. La Molina, 36 pp.

(Prepared by T. A. de Icochea)

Rosellinia Black Rot

The disease is prevalent in the tropics, where temperate and moist climates are found during the growing season. Especially heavy yield losses, rivaling those of late blight, are reported from Costa Rica and Ecuador. The disease has also been reported from Bolivia, Colombia, Peru, and Chile.

Symptoms

Plants become stunted and wilted. Leaves yellow, and plants slowly die. Stems may be cankered. Roots and stolons may be partially or completely destroyed, dark colored, and covered by a mat of rough, loose, fast growing strands of a grayish white mycelium (Fig. 57A). Affected tubers are partially or completely covered by the loose fungus strands at harvest (Plate 39). A hard, dark brown carbonaceous mass soon forms in affected tissue under the white mycelium. When cut, tubers often show a band of striate projections growing inward from the tuber surface (Fig. 57C). Single plants or groups of plants may be attacked in the field. Infection expands from affected plants in all directions. When soil is removed, rhizomorphlike strands of the fungus are found to extend from one plant to another. Tubers frequently rot before harvest.

Causal Organism

Morphology of the mycelium and, especially, the characteristic swellings of the hyphae above the septa (Fig. 57B) are characteristic of the genus *Rosellinia*, which produces no known fruiting bodies and should be considered a member of the Mycelia sterilia.

Rosellinia black rot can be distinguished from stem rot because the fungal strands from *Rosellinia* may be present on all parts below soil level, whereas *Sclerotium rolfsii* affects only the parts close to the soil surface. *Rosellinia* does not form round sclerotia. It also differs from the perfect stage of *Rhizoctonia*

Fig. 56. Stem rot. Symptoms on tuber: **A,** radiating white mycelium; **B,** later stage with sclerotia. **C,** Clamp connections in hyphae of *Sclerotium rolfsii*. (A, Courtesy L. W. Nielsen; B, courtesy T. A. de Icochea; C, courtesy T. A. de Icochea and L. J. Turkensteen)

solani (Thanatephorus cucumeris), which forms a velvety white sheet in close contact with plant parts touching the soil.

Epidemiology
The disease occurs in level or moderately sloping fields where water accumulates. The fungus develops in warm soils rich in organic matter at lower elevations, frequently on recently cleared land that has been forested or planted with pasture. It becomes a problem when potatoes are not rotated.

Other Hosts
The fungus affects carrots, beets, and members of the Brassicae and of the genera *Amaranthus, Rumex,* and *Polygonum.*

Control
1) Remove all debris from freshly cleared land.
2) Keep the land free of weeds, which may maintain the fungus.
3) Rotate potatoes with nonsusceptible crops.
4) Soil treatments (metam-sodium pentachloronitrobenzene) have reduced the disease to some extent.

Selected References

ORELLANA, H. A. 1978. Estudio de la enfermedad "Lanosa" de la papa en Ecuador. Fitopatologia 13:61–66.

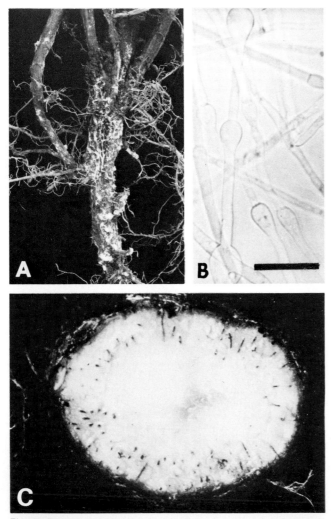

Fig. 57. Rosellinia black rot: **A,** infection of underground parts; **B,** characteristic swellings of mature hyphae (bar represents 20 µm); **C,** tuber rot with characteristic black lines projecting into tuber. (A, Courtesy E. R. French; B, courtesy G. deAbad and W. J. Hooker)

RODRIGUEZ, R. A. 1958. "Torbo," a tropical disease of potatoes. Plant Dis. Rep. 42:972–980.

(Prepared by L. J. Turkensteen)

Rhizopus Soft Rot

This fungus occurs throughout the world, but the disease is important chiefly in the tropics. Tuber losses under high temperatures may be high.

Symptoms
Water-soaked lesions on skin, initially small, enlarge rapidly and become a soft, watery rot extending into the flesh. Rotted tissue later becomes brown to chocolate brown. Mycelium in tissue is at first white and later dark. Grayish, later dark, sporangiophores develop on the surface. Rotted tissue develops zonate markings and at low relative humidity becomes a dry rot.

Causal Organisms
Rhizopus stolonifer (Fr.) Lind., *R. arrhizus,* and other *Rhizopus* spp. are typically saprophytes but may also be wound parasites on a wide range of fleshy storage organs of fruits and vegetables. Invasion is characterized by dissolution of the cell wall middle lamella. Germinating spores and mycelium growth are markedly inhibited by low temperature. The disease is most severe at temperatures of 20–40° C.

Control
1) Chilling tubers just after harvest to 2.5° C or below inactivates spore germination and mycelial growth.
2) Tuber treatment with disinfectants has been successful.
3) Avoid wounding tubers.

Selected References

MATSUMATO, T. T., and J. F. SOMMER. 1967. Sensitivity of *Rhizopus stolonifer* to chilling. Phytopathology 57:881–884.
LINK, G. K. K., and G. B. RAMSEY. 1932. Market diseases of fruits and vegetables. Potatoes. U.S. Dept. Agric. Misc. Publ. 98. 62 pp.
THAKUR, D. P., and V. V. CHENULU. 1974. Chemical control of soft rot of potato tubers caused by *Rhizopus arrhizus.* Indian Phytopathol. 27:375–378.

(Prepared by S. K. Bhattacharyya and R. Dwivedi)

Rhizoctonia Canker (Black Scurf)

Rhizoctonia canker, commonly called black scurf, is present in all potato-growing areas.

Symptoms
Black or dark brown sclerotia develop on surfaces of mature tubers (Plate 40). Sclerotia may be flat and superficial or large, irregular lumps resembling soil that will not wash off. The tuber periderm under such sclerotia is usually unaffected. Other tuber symptoms include cracking, malformation, pitting, and stem end necrosis.

Plants are most severely damaged in the spring shortly after planting; killing of underground sprouts delays emergence, especially in cold, wet soils. This results in poor, uneven stands of weak plants and subsequent yield reduction. Emerging potato sprouts may also be infected with cankers on the developing stem, often causing girdling and stem collapse (Fig. 58). Partial or complete girdling may promote a variety of plant symptoms, including stunting and rosetting of plant tops, cortical necrosis of woody stems, purple pigmentation of leaves, aerial tubers, upward leafroll, and often chlorosis, most severe at the top of the plant.

Reddish brown lesions on stolons cause stolon pruning or

tuber malformation. Roots are also pruned, resulting in a sparse root system.

The sexual (perfect) stage of the pathogen occurs on stems just above the soil line as a whitish gray mat on which basidiospores are formed, giving the surface a powdery appearance (Fig. 59A). The mat is easily rubbed off, and the stem tissue below the mat is healthy. These mats are often located above a lesion on the belowground portion of the stem (Plate 41).

A type of tuber malformation (Fig. 60), incompletely understood and not directly linked to *Rhizoctonia* infection, is frequent when *Rhizoctonia* is severe on tubers. The condition is believed to follow mycelial infection of the tip of very young tubers. Growth is retarded under the area of infection, and the tuber is deformed, often with superficial scalelike discolored tissue.

Causal Organism

The pathogen in its imperfect stage is *Rhizoctonia solani* Kühn and in its perfect stage (Fig. 59 C and D) *Thanatephorus cucumeris* Frank.) Donk. (syn. *Corticium vagum* Berk. & Curt., *Pellicularia filamentosa* (Pat.) Rogers, and *Hypochnus solani* Prill. & Delacr.). The *Rhizoctonia* hyphae are capable of anastomosis (hyphal fusion), and isolates have been further classified according to anastomosis groups. Isolates pathogenic to potato are generally placed in group AG-3 (Parameter et al).

The mycelium is generally tan to dark brown and hyphae are rather large (generally 8–10 μm in diameter). Young, vegetative hyphae have multinucleate cells and branch near the distal septum of a cell. Right-angle branching, constriction of branch hyphae at the point of origin, formation of a septum in the branch near the origin (Fig. 59B), and a prominent septal pore apparatus are all characteristics of *R. solani*.

Rhizoctonia produces a growth-regulating toxin that may be partially responsible for tuber malformation.

Disease Cycle

The pathogen overwinters as sclerotia on tubers, in soil, or as mycelium on plant debris in the soil. In the spring, when conditions are generally favorable, sclerotia germinate and invade potato stems or emerging sprouts, especially through wounds. Roots and stolons are invaded as they develop throughout the growing season. Sclerotial formation on new tubers is initiated at any time, depending on environmental

conditions; however, maximum development occurs as tubers remain in the soil after death of vines.

Epidemiology

Rhizoctonia populations may increase in soils where little or no rotation is practiced. Planting seed tubers that are heavily infested with sclerotia also favors inoculum buildup in soils. Environmental conditions favoring the pathogen are low soil temperatures and high moisture levels. The optimum soil temperature for disease development is 18° C and disease

Fig. 58. *Rhizoctonia* cankers on young stems, sprout girdling and death, and lateral sprouts forming from nodes below lesion. (Courtesy R. C. Rowe)

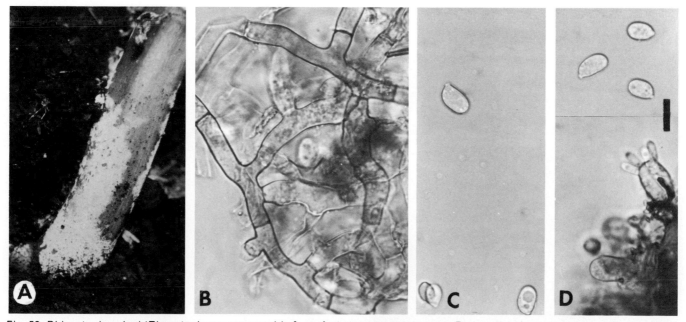

Fig. 59. *Rhizoctonia solani* (*Thanatephorus cucumeris*): **A**, perfect stage on potato stem; **B**, mycelial branching from such superficial mycelium on stem surface; **C**, basidiospores; **D**, basidium from potato stem surface near the soil line. Bar represents 10 μm and applies to B–D.

development decreases with increasing temperatures. High moisture levels in soils, especially those poorly drained, also tend to increase severity of sclerotial formation on new tubers.

Tuberborne sclerotia range in pathogenicity to stems and stolons from avirulence through moderate to high virulence. The influence of tuberborne sclerotia on the health of the following crop is not consistent and varies from essentially no deleterious effect to a measurable increase in sprout-pruning, stem cankers, and yield reduction.

High resistance within potato has not yet been identified.

Other Hosts

R. solani is a pathogen of numerous crops and weed hosts throughout the world. Its selective pathogenicity depends on the strain present.

Control

1) Seed treatment is not effective in heavily infested soils. Use disease-free seed combined with seed treatments such as the systemic fungicides (benomyl, thiabendazole, or carboxin) or, where acceptable, organic mercury.

2) Soil treatments of benomyl or pentachloronitrobenzene reduce soilborne inoculum, but the returns may not justify the cost.

Selected References

BIEHN, W. L. 1969. Evaluation of seed and soil treatments for control of Rhizoctonia scurf and Verticillium wilt of potato. Plant Dis. Rep. 53:425–427.

FRANK, J. A., and S. K. FRANCIS. 1976. The effect of a *Rhizoctonia solani* phytotoxin on potatoes. Can. J. Bot. 54:2536–2540.

HIDE, G. A., J. M. HIRST, and O. J. STEDMAN. 1973. Effects of black scurf (*Rhizoctonia solani*) on potatoes. Ann. Appl. Biol. 74:139–148.

JAMES, W. C., and A. R. McKENZIE. 1972. The effect of tuber borne sclerotia of *Rhizoctonia solani* Kühn on the potato crop. Am. Potato J. 49:296–301.

PARMETER, J. R., ed. 1970. *Rhizoctonia solani:* Biology and Pathology. Univ. of Calif. Press, Berkeley. 255 pp.

PARMETER, J. R., Jr., R. T. SHERWOOD, and W. D. PLATT. 1969. Anastomosis grouping among isolates of *Thanatephorus cucumeris.* Phytopathology 59:1270–1278.

SANFORD, G. B. 1956. Factors influencing formation of sclerotia by *Rhizoctonia solani.* Phytopathology 46:281–284.

VAN EMDEN, J. H., 1958. Control of *Rhizoctonia solani* Kühn in potatoes by disinfection of seed tubers and by chemical treatment of the soil. Eur. Potato J. 1:52–64.

VAN EMDEN, J. H., R. E. LABURYERE, and G. M. TICHELAAR. 1966. On the control of *Rhizoctonia solani* in seed potato cultivation

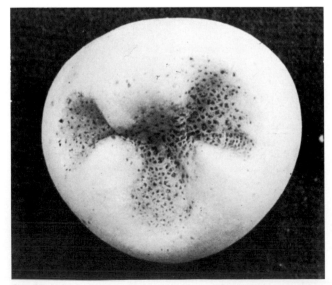

Fig. 60. Injury at tuber apex, common under certain conditions when *Rhizoctonia* is present. (Courtesy H. Torres)

in the Netherlands. Instituut voor Plantenziektenkundig Onderzoek. Mededeling 412. 42 pp.

WENHAM, H. T., B. L. MacKINTOSH, and H. A. BOLKAN. 1976. Evaluation of fungicides for control of potato black scurf disease. N. Z. J. Exp. Agric. 4:97–101.

(Prepared by J. A. Frank)

Violet Root Rot

Although infrequent, violet root rot has been reported from most of the major potato-growing areas of the world.

Symptoms

Aboveground symptoms are not distinctive. The foliage may become chlorotic, and plants may wilt and die suddenly in localized areas in the field. Belowground plant parts are often covered only with a reddish-purple mycelial network on the uninjured skin. Under the mycelial mats, tubers may have dark gray, somewhat sunken spots covered with purplish black sclerotia. The fungus tends to be limited to the cells near the periderm of the tuber. Wet rot of tubers may develop under mycelial mats.

Causal Organism

Helicobasidium purpureum (Tul.) Pat. (syn. *Rhizoctonia crocorum* (Pers.) DC) has young hyphae that are light violet, becoming more intensely violet with age. Hyphal branches arise at right angles close to a septum. The mycelium is branched, septate, and distributed evenly over the host surface. On occasion, strands are clearly visible. Dark brown to purplish black sclerotia form on the fungal mats. Sclerotia are essentially round, covered with a thick velvety felt, and vary in diameter from a few millimeters to several centimeters. The basidial stage is present infrequently on the base of potato stems near the soil surface as a white, superficial growth similar to that of *R. solani.* The basidium is hyaline, with two to four cells, each bearing a sterigma (10–35 μm in length) that produces a basidiospore (10–12 $\times$ 6–7 μm).

Disease Cycle

The fungus overwinters in the soil as sclerotia. These germinate in the spring and infect the crop. Basidiospores may spread the disease.

Other Hosts

The fungus parasitizes a wide range of hosts, the most important being carrot, lucerne (alfalfa), asparagus, and sugar beet.

Control

Rotation may be useful. Avoid rotation with other hosts.

Selected References

BUDDIN, W., and E. M. WAKEFIELD. 1927. Studies on *Rhizoctonia crocorum* (Pers.) DC. and *Helicobasidium purpureum* (Tul.) Pat. Trans. Br. Mycol. Soc. 12:116–140.

BUDDIN, W., and E. M. WAKEFIELD. 1929. Further notes on the connection between *Rhizoctonia crocorum* and *Helicobasidium purpureum.* Trans. Br. Mycol. Soc. 14:97–99.

KOTTE, W. 1930. Beobachtungen über den Parasitismus von *Rhizoctonia violacea* Tul. auf der Kartoffel. Ber. Dtsch. Bot. Ges. 48:43–51.

(Prepared by L. V. Busch)

Silver Scurf

Silver scurf is probably present in all of the major potato growing areas.

1. Giant-hill plants, taller than normal plants.

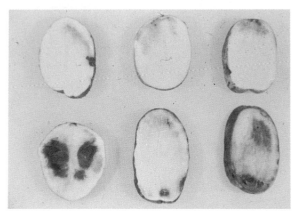

2. Mahogany browning in low temperature storage.

3. Low temperature leaf injury from temperatures above those freezing tissue.

4. Hail injury.

5. Wind injury.

6. Photochemical oxidant air pollution injury.

7. Air pollution injury from sulfur oxide. (Courtesy ASARCO Inc., Department of Environmental Sciences, Salt Lake City, UT)

8. Nitrogen (left) and phosphorus (right) deficiencies. (Courtesy Department of Soils and Plant Nutrition, University of California, Berkeley)

9. Potassium (left) and calcium (right) deficiencies. (Courtesy Department of Soils and Plant Nutrition. University of California, Berkeley)

10. Early sumptoms of magnesium deficiency (left) and manganese deficiency (right) (Left, Courtesy Department of Soils and Plant Nutrition, University of California, Berkeley; right, courtesy International Minerals and Chemicals Corp., Libertyville, IL)

11. Blackleg, *Erwinia carotovora* var. *atroseptica* infection in the field. (Courtesy M. D. Harrison)

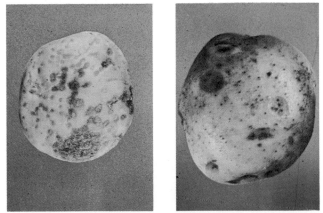

12. Bacterial soft rot. *Erwinia carotovora* lenticel infection of tubers.

13. Bacterial soft rot. *Erwinia carotovora* infection of tuber. (Courtesy J. E. Huguelet)

14. Brown rot. Plant infected with *Pseudomonas solanacearum*. (Courtesy E. R. French)

15. Brown rot. Tubers infected with *Pseudomonas solanacearum* exhibiting discolored eyes and vascular breakdown.

16. Ring rot. Interveinal chlorosis and upward curling of leaf margins. (Courtesy R. H. Larson)

17. Bacterial ring rot. Tuber infected with *Corynebacterium sepedonicum*. (Courtesy J. E. Huguelet)

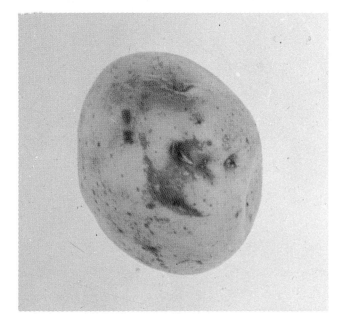

18. Pink eye. Infection at the tuber apex.

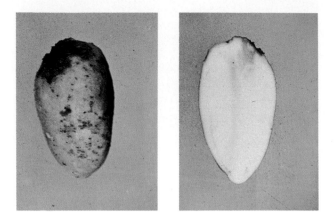

19. Pink eye. Tuber rot from severe pink eye infection.

20. Common scab. *Streptomyces scabies* infection of tubers ranging from surface russeting to deep lesions.

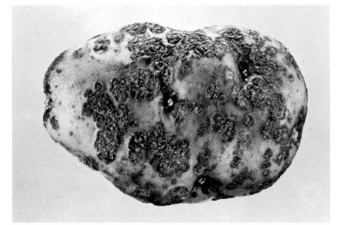

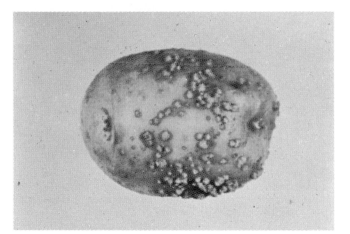

21. Deep common scab. *Streptomyces scabies* infection with sparse gray sporulation on the lesion surface. (Courtesy J. E. Huguelet)

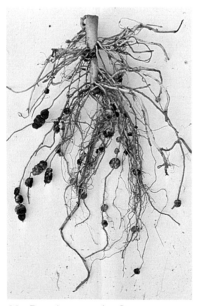

23. Powdery scab. *Spongospora subterranea* infection of potato roots. (Courtesy E. R. French)

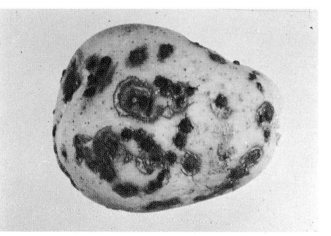

22. Powdery scab. Early and late infection of tuber. (Courtesy C. H. Lawrence and A. R. McKenzie)

25. Wart. *Synchytrium endobioticum* infection of tuber. (Courtesy M. C. Hampson)

24. Wart. *Synchytrium endobioticum* infection of meristems of stems, stolons, and tubers. (Courtesy R. Zachmann)

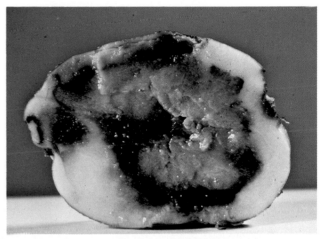

26. Leak. Tuber infected with *Pythium* sp. (Courtesy J. E. Huguelet)

27. Pink rot. Tubers infected with *Phytophthora erythroseptica*. (Courtesy R. C. Rowe)

28. Late blight. Rapidly expanding lesions with pronounced chlorotic border. (Courtesy D. P. Weingartner)

29. Late blight. Leaves and stems infected with *Phytophthora infestans.* Note sporulation on leaf surface. (Courtesy R. Zachmann)

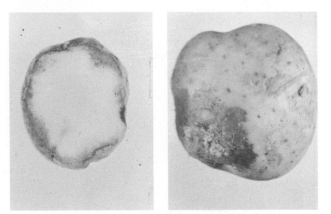

30. Late blight. Exterior and interior discoloration of tubers infected with *Phytophthora infestans.*

31. Powdery mildew. Recent infection of young leaves by *Erysiphe cichoracearum.* (Courtesy R. Zachmann)

32. Powdery mildew. Advanced stages of infection by *Erysiphe cichoracearum.* (Courtesy R. C. Rowe)

33. Early blight, *Alternaria solani* infection of leaves. (Courtesy L. J. Turkensteen)

34. *Pleospora herbarum* (*Stemphylium botryosum*). Infected leaves. (Courtesy L. J. Turkensteen)

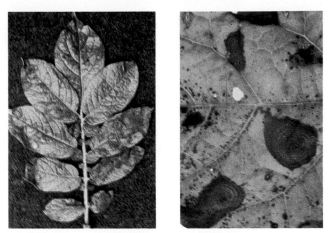

35. *Phoma andina.* Infected leaves. (Courtesy L. J. Turkensteen)

36. White mold. *Sclerotinia sclerotiorum* infection of stem. (Courtesy R. Zachmann)

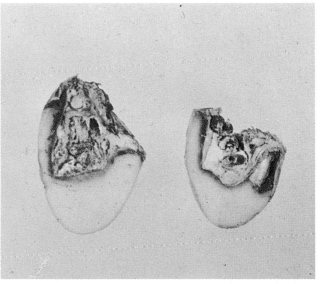

37. White mold. *Sclerotinia sclerotiorum* infection of tuber. (Courtesy T. de Icochea)

38. Stem rot. *Sclerotium rolfsii* stem infection at the soil line. (Courtesy L. J. Turkensteen)

39. Rosellinia black rot. Infection of tubers, with white mycelium on the tuber surface and soil. (Courtesy J. Bryan)

40. Rhizoctonia black scurf. Sclerotia of *R. solani* on tuber surface.

41. Rhizoctonia canker. *R. solani* infection of underground stems, sclerotia on seed tuber, and aboveground perfect stage (white area) on larger central stem. (Courtesy R. Zachmann)

42. Silver scurf. *Helminthospsorium solani* infection of tuber. (Courtesy J. E. Huguelet)

43. Black dot. *Colletotrichum atramentarium* on stem. (Courtesy E. R. French)

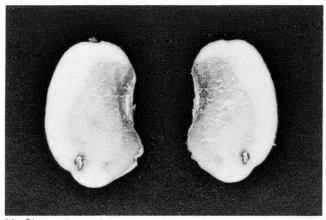

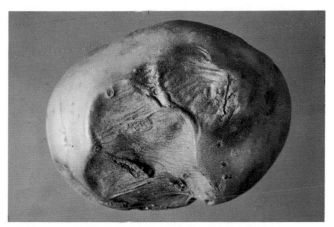

44. Charcoal rot. *Macrophomina phaseoli* infection of tuber. (Courtesy L. J. Turkensteen)

45. Gangrene. *Phoma exigua* var. *foveata* infection of tubers. (Courtesy R. Booth)

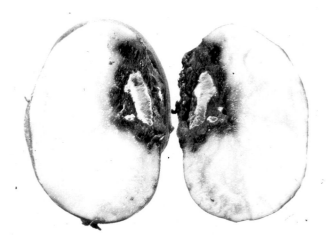

46. Fusarium dry rot infection of tuber (exterior view). (Courtesy J. E. Huguelet)

47. *Fusarium* dry rot infection of tuber (interior sections). (Courtesy J. E. Huguelet)

48. Fusarium dry rot. Cultural differences between *F. roseum* (fast growing pink) and *F. solani* (slow growing purple). (Courtesy L. W. Nielsen)

49. Fusarium wilt. Vine symptoms of *F. eumartii* infection.

50. Verticillium wilt of plant. (Courtesy R. Zachmann)

51. Verticillium wilt. Vascular discoloration in tubers.

52. Thecaphora smut. Infected tumors. (Courtesy J. Bryan)

53. Common rust. *Puccinia pittieriana* infection of leaves. (Courtesy E. R. French)

54. Deforming rust. *Aecidium cantense* infection of leaves and petioles. (Courtesy E. R. French)

55. Leafroll virus. Current season (primary) symptoms. (Courtesy R. Salzmann and E. R. Keller)

56. Leafroll virus. Secondary symptoms from tuberborne infection. (Courtesy E. R. French)

57. Leafroll virus symptoms. Marginal and interveinal chlorosis, in andigena type potato. (Courtesy R. A. C. Jones)

58. Rugose mosaic. Veinal necrosis and necrotic spotting of leaves.

59. Rugose mosaic. Mosaic mottle.

60. Rugose mosaic. Leaf drop and rugosity. (Courtesy J. Bryan)

61. Potato virus X. Mosaic mottle of this type is indistinguishable from that of other viruses such as PVY, PVM, PVS, etc. (Courtesy C. Fribourg)

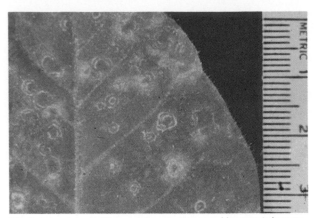

62. Potato virus X. Ring spot symptoms on Havana tobacco.

63. Potato virus M. Early symptoms.

64. Andean potato mottle virus. Secondary symptoms of severe patchy mottle and leaf deformation in the Peruvian cultivar, Revolucion. (Courtesy C. E. Fribourg)

65. Andean potato latent virus. Symptoms in Peruvian cultivar, Mi Peru. (Courtesy C. E. Fribourg)

66. Mop-top virus. Secondary leaf symptoms. (Courtesy L. Salazar)

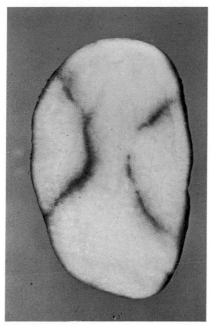

67. Mop-top virus. Primary tuber symptoms. (Courtesy R. A. C. Jones)

68. Mop-top virus. Secondary tuber symptoms (left) and healthy tuber (right). (Courtesy R. A. C. Jones)

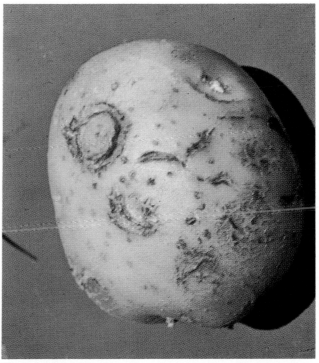

69. Tobacco rattle virus. Symptoms in leaves. (Courtesy D. P. Weingartner)

70. Tobacco rattle virus. Symptoms on tuber. (Courtesy D. P. Weingartner)

71. Potato yellow dwarf virus. (Courtesy S. Slack)

72. Alfalfa mosaic virus. (Courtesy I. Butzonitch)

73. Tobacco ringspot virus, (Andean potato calico). Early symptoms in Peruvian cultivar, Tichahuasi. (Courtesy C. E. Fribourg)

74. Andean potato calico. Advanced symptoms. (Courtesy C. E. Fribourg)

75. Potato yellow vein virus. (Courtesy J. Bryan)

76. Tomato spotted wilt virus. Symptoms in potato leaves. (Courtesy E. R. French)

77. Aster yellows mycoplasma symptoms. Rolling and pigmentation in upper leaves.

78. Aerial tuber in axils of leaves of plant with aster yellows mycoplasma symptoms. (Courtesy International Potato Center)

79. Hair sprouts from tuber infected with aster yellows mycoplasma (left) and sprouts from healthy tuber (right). (Courtesy L. J. Turkensteen)

80. Psyllid yellows. Result from toxins introduced during feeding by nymphs of the potato psyllid. (Courtesy E. Nelson)

81. Cysts of the golden nematode, *Globodera rostochiensis,* golden yellow before turning brown. (Courtesy International Potato Center)

82. Cysts of the nematode, *Globodera pallida,* white or cream colored before turning brown. (Courtesy International Potato Center)

83. Lesion nematode (*Pratylenchus penetrans*). Damage on tubers of Katahdin variety. (Courtesy W. F. Mai, B. B. Brodie, M. B. Harrison, and P. Jatala)

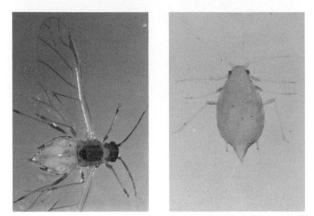

84. Buckthorn aphid (*Aphis nasturtii*). (Courtesy M. E. MacGillivray)

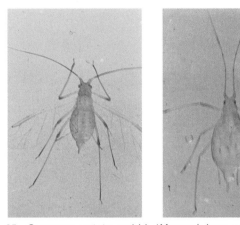

85. Common potato aphid (*Macrosiphum euphorbiae*). (Courtesy M. E. MacGillivray)

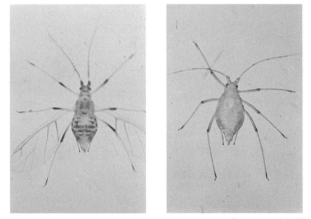

86. Foxglove aphid (*Aulacorthum solani*). (Courtesy M. E. MacGillivray)

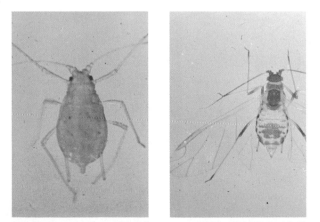

87. Green peach aphid (*Myzus persicae*). (Courtesy M. E. MacGillivray)

Symptoms

Small, localized, light brown, circular spots with indistinct borders frequently enlarge to cover a considerable area of the tuber. Affected areas have a distinct silvery sheen, particularly if the surface is wet (Plate 42). The color may deepen with age. If a large percentage of the surface is affected, tubers may shrivel during storage from excessive moisture loss. Red skinned varieties may lose their color.

Black dot and silver scurf produce similar blemishes on the tuber surface and may occur together. Margins of young silver scurf lesions are more definite and frequently have a sooty appearance caused by conidiophores and conidia. Silver scurf lesions do not have sclerotia.

Causal Organism

Helminthosporium solani Dur. & Mont. (syn. *Spondylocladium atrovirens* Harz.) has a hyaline mycelium that is septate, branched, and turns brown with age. Unbranched conidiophores are septate, with the conidia borne in whorls from the distal ends of the cells (Fig. 61). Spores are 7–8×18–64 μm, have up to eight septa, and are dark brown, rounded at the base, and pointed at the ends.

Disease Cycle

Transmission of the fungus is largely from infected seed pieces; soil transmission may also occur to a lesser extent. Infection takes place through the lenticels and periderm before tubers are dug, with intercellular and intracellular mycelium developing only in the periderm layer.

Epidemiology

High humidity is necessary for disease development. The longer mature tubers remain in the soil, the more severe the problem becomes. Minimum conditions for infection are 3° C and 90% rh. Disease continues to increase in storage, and further infection may develop if the tubers are kept at high relative humidity and temperature. Sporulation is more abundant on young lesions than on old ones.

Some varieties may be more susceptible than others.

Other Hosts

H. solani has never been found on any other host and only infects the tubers of potatoes.

Control

1) Use disease-free seed. Treat seed with benomyl.

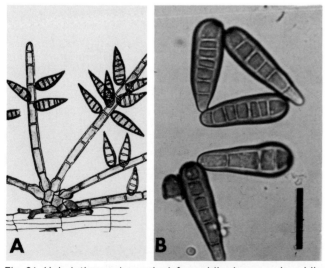

Fig. 61. *Helminthosporium solani*: **A,** conidiophores and conidia; **B,** conidia from tubers infected with silver scurf. Bar represents 20 μm. (A, Drawing from Taubenhaus, J. J. 1916. A contribution to our knowledge of silver scurf (*Spondylocladium atrovirens* Harz) of the potato. New York Bot. Gard. Memoirs 6:549–560.)

2) Harvest the tubers as soon as they are mature.

3) Ventilate storage area with warm air for drying, and store tubers at a low temperature consistent with wound healing and avoidance of other storage diseases.

Selected References

BOYD, A. E. W. 1972. Potato storage diseases. Rev. Plant Pathol. 51:297–321.

BURKE, O. D. 1938. The silver-scurf disease of potatoes. N.Y. Agric. Exp. Stn., Cornell, Bull. 692. 30 pp.

JELLIS, G. J., and G. S. TAYLOR. 1974. The relative importance of silver scurf and black dot: Two disfiguring diseases of potato tubers. Agric. Dev. Advis. Serv. Q. Rev. (London) 14:53–61.

JOUAN, B., J. M. LeMAIRE, P. PERENNEC, and M. SAILLY. 1974. Études sur la gale argentée de la pomme de terre, *Helminthosporium solani* Dur. et Mont. Ann. Phytopathol. 6:407–423.

SCHULTZ, E. S. 1916. Silver-scurf of the Irish potato caused by *Spondylocladium atrovirens*. J. Agric. Res. 6:339–350.

(Prepared by L. V. Busch)

Black Dot

The pathogen is common in many areas of the world, but the relative importance of the disease has not been well documented.

Symptoms

Black dot describes abundant, dotlike, black sclerotia on tubers, stolons, roots, and stems above and below ground (Fig. 62A, Plate 43). Symptoms vary from belowground rot of roots, stems, and stolons to aboveground yellowing and wilting of foliage. Foliage symptoms, which first occur at plant apices and later at mid and basal regions, may be confused with those of other wilt pathogens (e.g., *Verticillium* and *Fusarium* spp.). Lesions on belowground stems and stolons may also resemble Rhizoctonia disease of potato. Severe invasion of cortical tissue causes sloughing of the periderm. Following removal from the soil, roots may have a "stringy" appearance from decortication. As stems dry, cortical tissue is easily scaled away, an amethyst color is common inside the vascular cylinder, and sclerotia on stems develop abundantly externally and internally. With high relative humidity, the development of setae is inhibited.

Severe rotting of belowground plant parts and early death of the plant cause reduction in tuber size. At digging, pieces of dried stolons, with or without sclerotia, frequently adhere to the tubers. Stolons may be severed at any stage of tuber development, with the lesion usually 15–45 mm from the tuber. Sclerotia may develop on the upper surface of tubers and, in storage, grayish areas on tubers may closely resemble silver scurf.

Causal Organism

Colletotrichum atramentarium (Berk. & Br.) Taub. (syn. *C. coccodes* (Wallr.) Hughes) appears on a variety of media, including potato-dextrose agar, as a white, superficial mycelium. Sclerotia, 100 μm to 0.5 mm in diameter, are arranged in concentric rings and have acervuli that produce numerous spores and setae (Fig. 46B–D). Setae vary in length from 80 to 350 μm, are septate, and are pointed at the tip. Conidiophores develop free or in palisade layers and are subhyaline, 10–30 μm long, cylindrical, and tapering to slightly clavate. Spores en masse appear yellow to pink depending on media pH, are hyaline, 1–3 guttulate, attenuated at the basal end, round at the apical end, and 17.5–22 × 3.0–7.5 μm in size.

Histopathology

Mycelium and sclerotia are commonly associated with cortical and vascular tissue below the ground and at the base of the aboveground stem within several centimeters of the soil line. In certain instances, however, mycelium grows rapidly up the vascular cylinder of stems into leaves, even invading trichomes

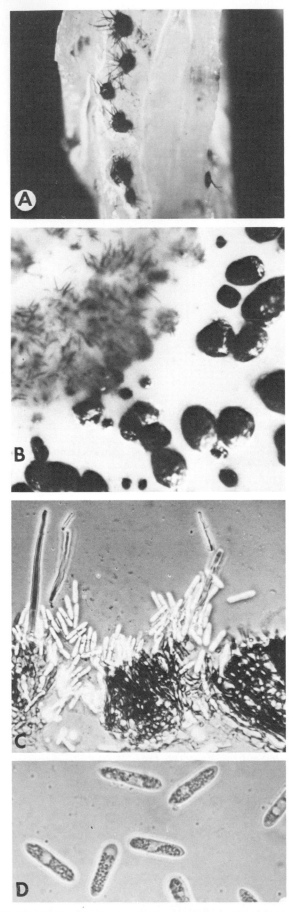

Fig. 62. Black dot: **A,** on potato root; **B,** portion of typical *Colletotrichum atramentarium* colony growing on nutrient Na-polygalacturonic acid medium; **C,** cross section of acervulus showing setae and conidia arising from acervulus; **D,** spores approximately 3.8 × 17.5 μm. (Courtesy J. R. Davis)

Epidemiology

Overwintering is by sclerotia on the surface of tubers or in plant debris in the field. The pathogen does not appear to be an active soil inhabitant, but it may survive in soil for long periods. *C. atramentarium,* generally regarded as a low grade pathogen that attacks under conditions of stress, commonly acts in combination with one or more additional pathogens, making its relative importance difficult to determine. Black dot is most frequently associated with light sandy soils, low nitrogen, high temperatures, and poor soil drainage.

Because *C. atramentarium* is elusive and lack of recognition is frequent, very little work on disease control has been done.

Other Hosts

In additional to potato, the fungus occurs on tomato and other plants in the Solanaceae (eggplant, pepper, tomato), on weed hosts such as *Physalis peruviana,* and, with inoculum increase, on *Datura stramonium,* a common weed.

Control

1) Clean seed, crop rotation, adequate fertility, and good irrigation management are commonly recommended.

2) No known potato cultivar offers resistance.

Selected References

DAVIS, J. R., and M. N. HOWARD. 1976. Presence of *Colletotrichum atramentarium* in Idaho and relation to Verticillium wilt (*Verticillium dahliae*). Am. Potato J. 53:397–398.

DICKSON, B. T. 1926. The "black dot" disease of potato. Phytopathology 16:23–41.

HARRISON, D. E. 1963. Black dot disease of potato. J. Agric. Victoria 61:573–576.

McINTYRE, G. A., and C. RUSANOWSKI. 1975. Scanning electron microscope observations of the development of sporophores of *Colletotrichum atramentarium* (B. et Br.) Taub. on infected potato periderm. Am. Potato J. 52:269–275.

STEVENSON, W. R., R. J. GREEN, and G. B. BERGESON. 1976. Occurrence and control of potato black dot root rot in Indiana. Plant Dis. Rep. 60:248–251.

THIRUMALACHAR, M. J. 1967. Pathogenicity of *Colletotrichum atramentarium* on some potato varieties. Am. Potato J. 44:241–244.

(Prepared by J. R. Davis)

Charcoal Rot

The fungus is worldwide but economically important only in warm regions where soil temperatures exceed 28° C.

Symptoms

Under hot conditions, the pathogen can attack potato stems and cause a sudden wilt and yellowing. Stem infection is not usually important. More important is tuber attack, which may occur before harvest and in storage, causing loss of the entire crop. Early symptoms develop around the eyes, near lenticels (particularly those that have enlarged), and frequently at the stolon attachment. The skin appears unaffected at first, with underlying tissue, usually that within 1 cm of the surface, becoming slightly water-soaked and light gray. Cavities filled with black mycelium and sclerotia form later. Rapidly invaded tubers, when cut, exhibit semiwatery, flabby breakdown, with color changing from yellowish (Plate 44) to pinkish to brown and finally to black. Wet rot may later develop from secondary invaders.

Causal Organism

The pathogen is *Macrophomina phaseoli* (Maubl.) Ashby (syn. *M. phaseolina* (Tassi) Goid., *Sclerotium bataticola* Taub.). Sclerotia within roots, stems, leaves, or fruits are black, smooth, hard, and 0.1–1 mm in diameter. They are smaller in culture. Pycnidia, dark brown on leaves and stems, are 100–200 μm in diameter. Single-celled conidia are hyaline, ellipsoid to obovoid, and $14{-}30 \times 5{-}10\ \mu$m. Pycnidia production in culture is rare except on propylene oxide sterilized leaf tissue in agar.

Epidemiology

The fungus maintains itself saprophytically on unthrifty or senescent plant parts and survives unfavorable periods as microsclerotia. Pycnidiospores are relatively short-lived. Tubers are infected through wounds, eyes, enlarged lenticels, and the stolon.

Tubers are predisposed to infection at temperatures of 32° C or higher. Rot development is restricted at low temperatures, slow at 20–25° C, and most rapid at 36° C and above. No secondary spread is apparent during storage, but infected tubers rot in warm storage. Rot stops in refrigerated storage, but when tubers are returned to warm temperature, the rot continues. Thus, seed from cold storage should be warmed before being planted so that infected tubers may be removed.

Most commercial cultivars are equally susceptible. Resistance exists in certain *Solanum chacoense* clones, in some of its hybrids, and in hybrids of the series Commersoniana.

Other Hosts

The fungus has been found on underground parts of an extremely wide range of plants, both cultivated and wild.

Control

1) Harvest early, before soil temperatures become high.

2) Avoid bruising and wounding of tubers in harvest and postharvest handling.

3) Field irrigation may be useful to prevent excessive soil temperature.

4) Do not leave tubers in soil after plants have matured.

5) Do not harvest during periods in which soil temperatures exceed 28° C.

6) Do not store tubers at high temperatures.

7) Do not use seed originating from areas where the disease is frequent.

Selected References

BHARGAVA, S. N. 1965. Studies on charcoal rot of potato. Phytopathol. Z. 53:35–44.

GOTH, R. W., and S. A. OSTAZESKI. 1965. Sporulation of *Macrophomina phaseoli* on propylene oxide-sterilized leaf tissues. Phytopathology 55:1156.

HOLLIDAY, P., and E. PUNITHALINGAM. 1970. *Macrophomina phaseolina*. No. 275 in: Descriptions of Pathogenic Fungi and Bacteria. Commonw. Mycol. Inst., Kew, Surrey, England. 2 pp.

PUSHKARNATH. 1976. Potato in Sub-Tropics. Orient Longman, Ltd., New Delhi, 289 pp.

SAHAI, D., B. L. DUTT, and K. D. PAHARIA. 1970. Reaction of some wild and cultivated potato varieties to charcoal rot. Am. Potato J. 47:427–429.

THIRUMALACHAR, M. J. 1955. Incidence of charcoal rot of potato in Bihar (India) in relation to cultural conditions. Phytopathology 45:91–93.

von AMANN, M. 1960. Untersuchungen über einen sklerotienbildenden Pilz an Kartoffeln, vermutlich *Sclerotium bataticola* (Taub.), synonym *Macrophomina phaseoli* (Maubl.) Ashby. Z. Pflanzenkr. Pflanzenschutz 67:655–662.

(Prepared by L. J. Turkensteen and W. J. Hooker)

Gangrene

The pathogen *Phoma exigua* var. *foveata* was first described in 1940 and is now prevalent in most northern European countries and parts of Australia. *P. exigua* var. *exigua* occurs in most European countries, Russia, the United States, Canada, and Australasia.

Symptoms

Small dark depressions develop in the tuber skin, usually at wounds, eyes, or lenticels, and may enlarge to form "thumbmark" or larger, irregularly shaped, sharp-edged lesions, the surface area of which is often unrelated to rot depth. Internally, diseased tissue is well defined. Rots caused by *Phoma exigua* var. *foveata* are usually extensive and dark brown or purplish (Plate 45), with variously shaped cavities; those caused by *P. exigua* var. *exigua* are smaller, become restricted, and are usually black with small cavities. Pycnidia may form singly or in clusters on lesions or in the mycelium that lines cavities. Infrequently, lesions may be only of skin thickness, becoming extensive, dark, and irregularly shaped; this condition is termed skin necrosis.

Causal Organism

Either of two varieties of *Phoma exigua* Desm. may cause gangrene. The principal cause is *P. exigua* var. *foveata* (Foister) Boerema (syn. *P. foveata* Foister; *P. solanicola* f. *foveata* (Foister) Malcolmson; *P. exigua* Desm. f. sp. *foveata* (Foister) Malcolmson & Gray). The more ubiquitous but weaker parasite is *P. exigua* Desm. var. *exigua* (syn. *P. solanicola* Prillieux & Delacroix; *P. tuberosa* Melhus, Rosenbaum, and Schultz; *P. exigua* Desm. f. sp. *exigua* Malcolmson and Gray).

The two fungi have similar morphological characteristics. Pycnidia are usually globoid (90–200 μm) and dark brown to black. Initially subepidermal, they become erumpent and extrude hyaline, nonseptate, cylindrical pycnidiospores (4–5 $\times$ 2–3 μm). In culture on 2% malt agar, *P. exigua* var. *foveata* (having nonzonate colonies) is readily distinguished from *P. exigua* var. *exigua* (having zonate colonies) (Fig. 63) by its production of anthraquinone pigments that turn red within seconds on exposure to ammonia vapor.

Disease Cycle

Infected or contaminated seed tubers produce diseased stems, in which infection remains latent during the growing season unless the stems become moribund. Pycnidia appear in sporadic groups, usually associated with nodes, as stems begin to senesce either naturally or through chemical desiccation. Raindrops wash pycnidiospores into the soil and spread inoculum to neighboring plants. Rots in mother tubers usually continue active in the soil, produce pycnidia, and constitute another important source of inoculum for tubers at harvest. Before harvest, tuber infection may occur through eyes and proliferated lenticels, usually when soil moisture is high. Most gangrene, however, develops after harvest through damage to the tuber skin. Wounding introduces infection from contaminated soil on the tuber surface or stimulates development of the fungus already latent in the periderm. Wound infection may occur at lifting, grading, or at any time during handling.

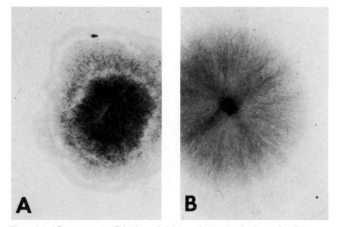

Fig. 63. Gangrene. Distinguishing characteristics: **A,** *Phoma exigua* var. *exigua,* zonate culture; **B,** *Phoma exigua* var. *foveata,* nonzonate culture, on malt agar. (Courtesy C. Logan; photographs by G. Little)

Epidemiology

In the United Kingdom, higher than average soil temperatures (20–24° C) during crop growth keep *P. exigua* var. *foveata* inoculum levels low. However, wet soil conditions, night frosts, or low day temperatures (less than 12° C) around harvest time and low storage temperatures (2–10° C) following lifting, grading, or handling all increase gangrene incidence. Delays in haulm destruction of seed crops or in harvest of both seed and market crops encourage inoculum buildup in the soil. In field soils in the absence of a potato host, *P. exigua* var. *foveata* usually becomes undetectable by present isolation or baiting techniques after 18 months, whereas *P. exigua* var. *exigua* is detectable in most soils.

Other Hosts

P. exigua var. *exigua* occurs on various parts of a wide range of plants. *P. exigua* var. *foveata* occurs mostly on potato but has occasionally been found on weed species growing in potato fields.

Control

1) Avoid highly susceptible cultivars, damage to the tuber skin, and exposure to low temperatures, especially after damage.

2) Burn vines and harvest tubers as soon as practical. Hold tubers at 18–20° C for one week to permit wound healing.

3) Disinfect tubers with organic mercury dips (where permitted), by fumigation with 2-aminobutane, or by mist sprays of thiabendazole within three weeks of harvest.

Selected References

BOEREMA, G. H. 1976. The *Phoma* species studied in culture by Dr. R. W. G. Dennis. Br. Mycol. Soc. Trans. 67:289–319.

BOYD, A. E. W. 1972. Potato storage diseases. Rev. Plant Pathol. 51:297–321.

LOGAN, C., R. B. COPELAND, and G. LITTLE. 1975. Potato gangrene control by ultra low volume sprays of thiabendazole. Ann. Appl. Biol. 80:199–204.

(Prepared by C. Logan)

Fusarium Dry Rots

Fusarium dry rots are found on potatoes worldwide.

Symptoms

This disease affects tubers in storage and planted seed tubers.

After about one month of storage, tuber lesions at wounds are visible as small brown areas. Infection slowly enlarges, and periderm over the lesion sinks and wrinkles, sometimes in concentric rings, as the dead tissues dry out (Fig. 64). Fungus pustules containing mycelia and spores may emerge from the dead periderm. Rotted tubers shrivel and become mummified (Plate 46).

Internal necrotic areas are shades of brown from fawn to dark chocolate, with the advancing margin faint for lighter shades and distinct for darker shades. Older dead tissues assume a variety of colors, develop cavities lined with mycelia and spores, and are dry and punky in texture (Plate 47).

When relative humidity in storage is saturated or approaching saturation, *Erwinia* spp. are frequently secondary

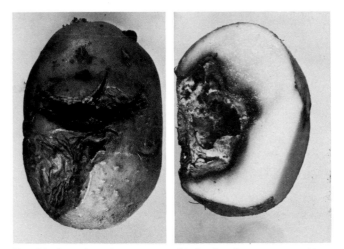

Fig. 64. *Fusarium* dry rot wound infection from storage. Note periderm wrinkling over rotted tissue and internal cavity. (Courtesy L. W. Nielsen)

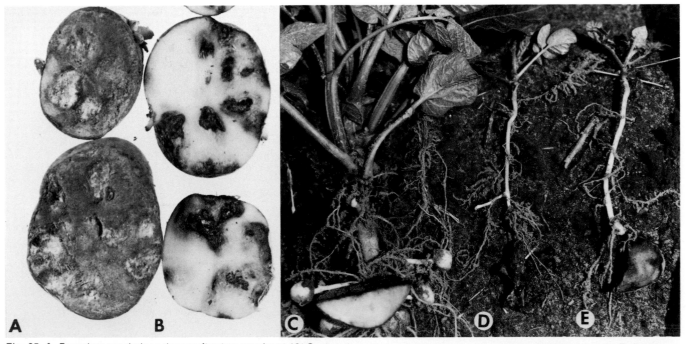

Fig. 65. **A,** *Fusarium* seed piece decay after two weeks at 13° C, showing pits on cut surfaces and mycelium growing in pits. **B,** Periderm removed from surfaces exposes rot from several infections. Comparative plant growth from healthy seed piece (**C**) and *Fusarium* decayed seed pieces (**D** and **E**). (Courtesy L. W. Nielsen)

invaders through the *Fusarium* lesions and rapidly rot the remainder of the tuber. Suspended bacteria and juices exuded from the soft rot endanger surrounding tubers.

Whole-tuber seed becomes infected through wounds during storage or preparation for planting. The cut surfaces of large seed tubers are major infection courts. In stored seed tubers, brown to black flecks appear on the cut surface in about one week and depressions or pits form in two weeks (Fig. 65A and B). Mycelia often grow on the depressed surfaces, and under humid conditions, depressions may become slimy and black from bacterial growth. Soft rotting bacteria may also invade through the *Fusarium* lesions and accelerate decay. With numerous cut-surface infections, lesions coalesce; the seed piece rots from the surface inward; and buds (eyes) are destroyed as decay progresses.

In the field, the shriveling of infected seed tubers and pitting of infected pieces may not be evident. The surface over the lesions is brown, and the underlying necrotic tissues have fewer cavities. Necrotic tissue may attract soil insects and larvae such as the seed-corn maggot, which is a vector of *Erwinia* species. In wet soils, these species often enter as secondary pathogens. *Fusarium* spp. alone or in conjunction with *Erwinia* spp. partially or completely destroy the seed piece, resulting in extreme variability in plant size and many missing plants (Fig. 65D and E). Often single sprouts emerge; these are small, grow slowly, produce few marketable tubers, and have a high incidence of blackleg.

Causal Organisms

Fusarium solani (Mart.) App. & Wr. emend Snyd. & Hans. 'Coeruleum' and *F. roseum* (Lk.) Snyd. & Hans. 'Sambucinum' are most frequently implicated in seed piece decay. In some regions one species is dominant over the other, but both are often associated with a seed stock. *F. solani* is most frequently encountered and is the more aggressive pathogen. Both grow and are maintained on potato-dextrose agar, and the acidified medium facilitates their isolation when bacteria are present. *F. roseum* grows more rapidly, forming a thin, white mycelial mat and abundant pink to salmon spores. The slower growing *F. solani* forms a denser, white mycelial mat that exhibits a purple pigment with age (Plate 48). *F. solani* sporulates in culture more sparsely than does *F. roseum* (Fig. 66). The optimum temperature for growth in culture is 20–25° C and for infection, 10–20° C.

F. roseum 'Avenaceum' also causes a dry rot of potatoes, but less frequently than the other species do.

Histopathology

Fusarium spp. cannot infect intact tuber periderm or lenticels. Cuts and periderm-breaking wounds incident to harvesting, storage, grading, and transport are the major infection courts. Wounds from insect and rodent feeding and frost are sometimes infected. The Fusaria can also invade surface lesions of powdery scab, late blight, mop top virus, and possibly other diseases.

Hyphae are at first intercellular, becoming intracellular in dead cells. In spreading lesions, hyphae may be sparse in intercellular spaces, with host cells showing little reaction to the fungus. Toward the center of the lesion, less starch is present, and the mycelium, usually abundant, may be confined to the intercellular spaces by suberin deposited in host cell walls and intercellular spaces. In susceptible tissue, starch hydrolysis and suberin deposition are lacking. Small lesions restricted near the site of infection may be underlaid by a continuous layer of wound meristem cells with suberin deposition. With other isolates, hyphae kill and penetrate cells within two cells of normal-appearing tissue. Details of the reaction depend on the pathogen, the resistance of the tuber, and the part of the lesion examined.

Disease Cycle

Fusarium spp. can survive for several years in field soil, but the primary inoculum is generally borne on seed tuber surfaces. Surfaceborne propagules contaminate containers and equipment used in handling or storing potatoes and enter wounds incident to handling seed tubers. Infected seed tubers and pieces decay and infest the soil that adheres to the surfaces of harvested tubers.

Epidemiology

Tubers of potato cultivars differ in susceptibility to *F. solani* and *F. roseum*, but none tested was immune to either pathogen. Certain cultivars are tolerant to both.

Tubers are tolerant to infection when harvested. Susceptibility increases during storage and reaches its maximum in early spring about planting time.

Wound healing can reduce infection. Deposition of suberin in the cell walls does not prevent infection, but wound periderm does. Wound periderm forms in three to four days at approximately 21° C with adequate aeration and humidity but more slowly at lower temperatures. At 15° C, near optimum for infection, a period of approximately eight days is required to form periderm; wound healing is not effective at this or lower temperatures.

Dry rot develops most rapidly in high relative humidity and at 15–20° C. Relative humidities about 70% do not alter rot development, but lower humidities retard infection and disease development. Disease development continues at the coldest temperatures safe for potatoes.

If the soil temperature and moisture are suitable for rapid sprout growth and emergence, seed tuber or piece decay after planting may be of little consequence. Conditioning seed tubers from cold storage at 20–25° C for one week before cutting pieces reduces decay and accelerates sprout growth. Holding contaminated cut seed several days or weeks before planting or planting in soils too cold or dry for prompt sprout emergence and plant growth will accentuate losses. Excessively wet soils after planting increase secondary infection by *Erwinia* spp.

Other Hosts

F. solani isolates from *Colocasia* corms can infect potato tubers. Generally, tuber-rotting Fusaria do not infect other plants or plant organs.

Resistance

Differences in resistance exist among potato cultivars. Relative ranking of resistance is influenced by the *Fusarium* sp. used for inoculation. Seed lots infected with potato virus X are relatively more resistant than those free from the virus when harvested within three weeks of top kill.

Control

1) For storage and seed purposes, harvest tubers from dead vines.

2) Use all precautions with machinery and equipment to

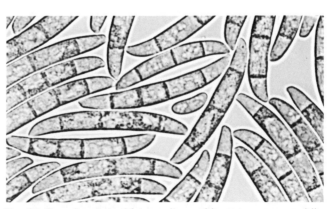

Fig. 66. *Fusarium solani* from culture with many macroconidia and a single microconidium in center. (×950) (Courtesy P. E. Nelson)

prevent wounding during harvest and storage.

3) Provide high humidity and good ventilation early in storage to facilitate wound healing, and provide aeration during storage.

4) Seed tubers may be treated with a fungicide, dust, or liquid spray before storage.

5) Do not move stored tubers until they are ready for planting.

6) Warm seed tubers from cold storage to 20–25° C for a week before planting or cutting pieces.

7) Plant seed immediately after cutting in soils sufficiently warm and moist to promote prompt sprout growth and good wound healing.

8) Spray or dip seed tubers with fungicide suspensions or treat pieces with 7–8% fungicidal dusts.

9) Handle treated seed with noncontaminated containers and equipment.

Selected References

BOYD, A. E. W. 1952. Dry-rot disease of the potato. Ann. Appl. Biol. 39:322–357.

CUNNINGHAM, H. S., and O. A. REINKING. 1946. Fusarium seedpiece decay of potato on Long Island and its control. N.Y. Agric. Exp. Stn., Geneva, Bull. 721. 32 pp.

JONES, E. D., and J. M. MULLEN. 1974. The effect of potato virus X on susceptibility of potato tubers to *Fusarium roseum* 'Avenaceum.' Am. Potato J. 51:209–215.

LEACH, S. S., and L. W. NIELSEN. 1975. Elimination of fusarial contamination on seed potatoes. Am. Potato J. 52:211–218.

McKEF, R. K. 1954. Dry-rot disease of the potato. VIII. A study of the pathogenicity of *Fusarium caeruleum* (Lib.) Sacc. and *Fusarium avenaceum* (Fr.) Sacc. Ann. Appl. Biol. 41:417–434.

NIELSEN, L. W. 1949. Fusarium seedpiece decay of potatoes in Idaho and its relation to blackleg. Idaho Agric. Exp. Stn. Res. Bull. 15. 31 pp.

NIELSEN, L. W., and J. T. JOHNSON. 1972. Seed potato contamination with Fusarial propagules and their removal by washing. Am. Potato J. 49:391–396.

SMALL, T. 1944. Dry rot of potato (*Fusarium caeruleum* (Lib.) Sacc.).

Investigation on the sources and time of infection. Ann. Appl. Biol. 31:290–295.

(Prepared by L. W. Nielsen)

Fusarium Wilts

These diseases are widespread and most severe where potatoes are grown at relatively high temperatures or when seasons are hot and dry.

Symptoms

Several *Fusarium* pathogens cause essentially similar symptoms. Tubers exhibit surface blemishes and decay, including stem end browning and decay at the stolon attachment, and internal vascular discoloration that severely impairs market quality because such tubers cannot be removed during grading.

On vines, symptoms include cortical decay of roots and lower stems; vascular discoloration or rot in the lower stem; wilting; chlorosis, yellowing, or bronzing of foliage; rosetting and purpling of aerial parts; aerial tubers in leaf axils; and premature death of the plant. Additional symptoms vary with the pathogen involved and the environment.

Eumartii Wilt. Generally the most important and severe, this wilt becomes apparent toward the end of the growing season. The first symptom is yellowing between the veins of the youngest leaves, producing islands of green against a chlorotic background (Plate 49). Chlorotic areas later become necrotic. Affected leaves become yellowish bronze, wilt, dry, and hang on the stem, which eventually dies. Rolling and rosetting occur under moist conditions. Leaf discoloration and death may be most severe on one side of the stem or on stems on one side of the plant. Internally, the pith is often discolored at the nodes (Fig. 67A), even in those near the stem tip. Vascular tissues of the stem and leaf petioles are deep brown. The underground stem does not rot until later stages of disease.

Tubers are sunken at the stolon attachment, with brown

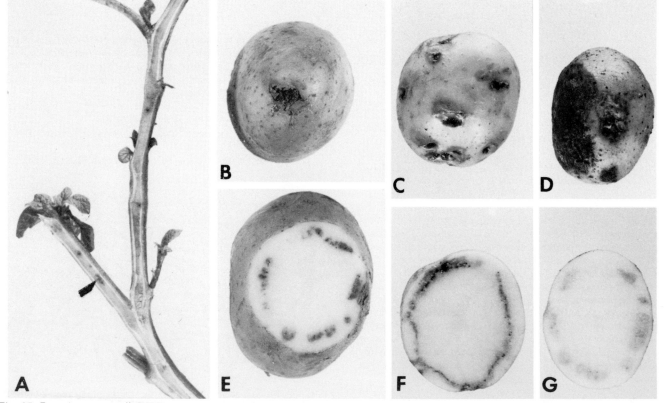

Fig. 67. *Fusarium eumartii:* **A,** pith necrosis (arrow) in stem; **B** and **D,** necrosis at stolon attachment; **C,** discoloration at eyes; **E** and **F,** internal vascular discoloration; **G,** water-soaked and firm vascular tissue.

necrosis extending into the tuber to various depths (Fig. 67B–G). Firm, brown circular lesions (up to 2.5 cm in diameter) may be present elsewhere on the tuber surface. In transverse section at the stem end, slight to severe vascular browning is evident, and the vascular ring may have a few thick, black strands (up to 1 mm in diameter) or more numerous, smaller brown to tan, netlike strands in the vascular ring. Highly diagnostic, but not present in every tuber, is a water-soaked, firm, light brown to tan discoloration extending 3–5 mm on either side of the vascular ring. This discolored area is firm, does not produce exudate as in bacterial ring rot or brown rot, and usually shows little tendency to break down with secondary rots. Vascular necrosis extending into the eyes may cause eyes to be brown and necrotic.

Oxysporum Wilt. Usually milder than eumartii wilt, this disease is a typical vascular wilt in contrast to the other Fusarium wilts described here, which are more nearly cortical rots. Symptoms appear during the middle of the growing season; wilt is rapid, giving the impression that the lower stem has been cut off; and the plant is prematurely killed. Yellowing begins at the lower leaves and progresses up the plant. Vascular discoloration of the stem is confined to portions below or slightly above the soil line. Tubers generally show discoloration of the vascular strands and usually no stem end rot.

Tuber infection through wounds or possibly lenticels causes circular lesions and a dry rot in storage. This condition has been associated with high humidity and temperature.

Avenaceum Wilt. This disease is comparable in severity to oxysporum wilt and generally less severe than eumartii wilt. It develops from mid to late season. Symptoms may be more severe on one side of the plant. Wilting and rapid collapse of the plant are common in hot, dry weather. A different response, possibly when growing conditions are more favorable, consists of chlorosis at bases of apical leaves, followed by general bunching of leaflets, chlorosis of the plant beginning at the base, shortening of internodes, carbohydrate accumulation in aboveground portions of the plant, red or purple pigmentation, and aerial tubers in leaf axils. The plant resembles those affected by mycoplasma or psyllid yellows. However, vascular discoloration may be seen in the lower portion of the stem up to six inches above the soil. Tip burn and loss of lower leaves is common. Early season infection produces severely dwarfed plants similar to those with yellow dwarf disease.

Tubers show dry stem end rot and vascular tissue that is discolored brown, lacks the water-soaked border of eumartii but is characteristically dry, and may be gray to pink.

F. solani Wilt. This wilt is distinct from eumartii wilt. It is characterized by rotting of the root system, the stem pith, and the lower and underground stem, with dry shredding of the strands of woody tissue, and wilting and yellowing of foliage. In moist conditions rosetting of tops and aerial tubers appear.

Tuber infection follows wound infection and differs from other wilts in lacking typical vascular discoloration. Neither stolons nor tubers are directly infected from the parent plant.

Causal Organisms

Taxonomy of genus *Fusarium* is complex. Species designations used here are those used in disease discriptions. Designation of species by the system of Snyder and Hansen is given in synonyms.

The four pathogens associated with the Fusarium wilts are: *F. eumartii* Carp. (syn. *F. solani* f. sp. *eumartii* (Carp.) Snyd. & Hans.); *F. oxysporum* Schl. (syn. *F. oxysporum* Schl. f. sp. *tuberosi* (Wr.) Snyd. & Hans.); *F. avenaceum* (Fr.) Sacc. (syn. *F. roseum* (Lk.) Snyd. & Hans.); and *F. solani* (Mart.) App. & Wr. (syn. *F. solani* f. sp. *eumartii* (Carp.) Snyd. & Hans.).

Descriptions on the basis of spore characteristics will not be attempted because of the well known variability of *Fusarium* as influenced by environmental factors.

Isolations of *F. eumartii* are readily obtained from roots and less readily from stems. Discolored tissue from upper stems and tuber apices is sterile. Isolation from stored tubers is difficult.

Tissue discoloration develops in advance of the fungus and is apparently associated with toxic substances.

F. oxysporum is easily isolated from roots and lower stems and, with difficulty, from stored tubers. Wilting is, in part, due to toxins.

F. avenaceum is successfully isolated from vascular discolored stem tissue below or close to the soil line and from discolored vascular tuber tissue.

F. solani is readily isolated from discolored stem tissue.

Histopathology

Diseases caused by the four wilt fungi are essentially similar. Root tips, following infection from the soil, become water-soaked. Epidermal cells of young roots are invaded. Cell walls become softened and swollen, and cortical necrosis follows. Xylem of roots and stems is invaded; vessels become plugged with granular material; and surrounding cells of outer phloem and cortex break down in ways varying with the particular pathogen involved.

Disease Cycle

Fusarium wilts are typically soilborne, and the disease is transmitted with varying degrees of effectiveness from inoculum within and on seed tubers.

F. eumartii survives in field soil for long periods without noticeable reduction in pathogenicity when potatoes are again planted in the field. Because of this disease, many fields have been abandoned for potato production. The other wilt organisms may be shorter-lived in the absence of potatoes, but evidence is lacking.

Planting potatoes in artificially infested soil or placing inoculum on freshly cut seed efficiently establishes the disease. Infection is through roots into the stem and, except for *F. solani,* from the stem through the stolon into the developing tubers.

Infected seed pieces with stem end rot transmit disease to the new plant, more efficiently with eumartii than with oxysporum. Inoculum is introduced into new fields primarily through the planting of infected seed tubers. Continual potato production, particularly replanting infected tubers, accelerates inoculum buildup. Inoculum is dispersed from infested fields by surface drainage water, windblown soil, soil carried on implements, etc.

Epidemiology

Wilts are most severe at high temperatures and particularly when plants are under stress in dry, hot growing conditions. Although evidence is lacking, the rosette symptom with aerial tubers probably follows increased availability of water and somewhat cooler temperatures.

F. eumartii is capable of infection at lower soil temperatures (20 and 24° C), whereas *F. oxysporum* and *F. avenaceum* are more pathogenic at 28° C. *F. solani* in culture can grow at 35° C but grows most rapidly at 30° C.

Other Hosts

The cultivated potato is the only known natural host for the several *Fusarium* spp. causing wilt of the crop. Morphologically similar pathogens attack plants of widely divergent types. Species differentiation has been based on pathogenicity specific to a particular plant species. Several *Solanum* spp. related to potato have been experimentally infected with *F. eumartii.*

Resistance

Minor differences in resistance within *S. tuberosum* are known, but identified resistance is not sufficiently high to be of general use. *S. spegazzinii, S. acaule,* and *S. kurtzianum* seedlings carry considerable resistance to root infection by *F. eumartii* in infested soil.

Control

1) Grow potatoes in land free from wilt fungi.
2) Tubers infected with Fusarium wilt should not be used for seed.

3) Avoid contamination of clean fields by inoculum transfer through infested soil or diseased tubers and plant refuse.

Selected References

GOSS, R. W. 1923. Relation of environment and other factors to potato wilt caused by *Fusarium oxysporum*. Neb. Agric. Exp. Stn. Res. Bull. 23. 84 pp.

GOSS, R. W. 1924. Potato wilt and stem-end rot caused by *Fusarium eumartii* Neb. Agric. Exp. Stn. Res. Bull. 27. 83 pp.

GOSS, R. W. 1940. A dry rot of potato stems caused by *Fusarium solani*. Phytopathology 30:160–165.

McLEAN, J. G., and J. C. WALKER. 1941. A comparison of *Fusarium avenaceum*, *F. oxysporum*, and *F. solani* var. *eumartii* in relation to potato wilt in Wisconsin. J. Agric. Res. 63:495–525.

RADTKE, W., and A. ESCANDE. 1975. Vergleichende Untersuchungen über verschiedene Methoden zur Inokulation von Kartoffelsämlingen mit *Fusarium solani* (Mart.) Sacc. f. sp. *eumartii* (Carp.) Snyder et Hansen. Potato Res. 18.243–255.

SNYDER, W. C., and H. N. HANSEN. 1940, 1941, 1945. The species concept in *Fusarium*. Am. J. Bot. 27:64–67; 28:738–742; 32:657–666.

UPSTONE, M. E. 1970. A corky rot of Jersey Royal potato tubers caused by *Fusarium oxysporum*. Plant Pathol. 19:165–167.

(Prepared by J. E. Huguelet and W. J. Hooker)

Verticillium Wilt

The disease apparently occurs wherever potatoes are grown, although it may be confused with other diseases that cause early maturity.

Symptoms

Verticillium wilt causes early senescence of plants. Leaves, which become pale green or yellow and die prematurely, are described as "early dying" or having "early maturity."

During the growing season, plants may lose their turgor and wilt, especially on hot, sunny days (Plate 50). Single stems or leaves on one side of the stem may wilt first. Vascular tissue of stems becomes a light brown, best observed if the stem is severed at about ground level with a long slanting cut. Externally visible necrotic stem streaking occurs in certain cultivars when soil moisture and fertility are high.

Tubers from infected plants, but not necessarily all tubers, usually develop a light brown discoloration in the vascular ring (Plate 51, Fig. 68A); severe vascular discoloration may extend over halfway through the tuber. Cavities may develop inside severely affected tubers. Pinkish or tan discoloration (see pink eye) may develop around the eyes or as irregular blotches on the surface of affected tubers. This may be confused with mild late blight infection.

Causal Organisms

Verticillium albo-atrum Reinke & Berth. develops septate, resting dark mycelium on stems in the field and also in culture, in contrast to *V. dahliae* Kleb., which forms dark mycelial strands with black, thick-walled pseudosclerotia (Fig. 69), also called microsclerotia, 30–60 μm in diameter. Vegetative hyphae of both are similar (2–4 μm in diameter and colorless). Conidiophores are septate with side branches (Fig. 68B), swollen at the base, and arranged in a whorl. First-formed conidia of *V. albo-atrum* are 6–12 × 2.5–3 μm. Those of *V. dahliae* are 3–5.5 × 1.5–2 μm. Conidia produced later, particularly those in culture, may be considerably smaller, 3–6 × 2–3 μm. Conidia are usually single celled but may be one-septate (Fig. 68C and D).

Both types may be present within a single potato plant.

Disease Cycle

Infection is through root hairs, wounds (including those at points of emergence of adventitious roots), and through sprout and leaf surfaces. Hyphae progress intracellularly and intercellularly to the xylem (Fig. 68D). Transport of conidia within vessels of potato is probable. Conidia are short-lived and

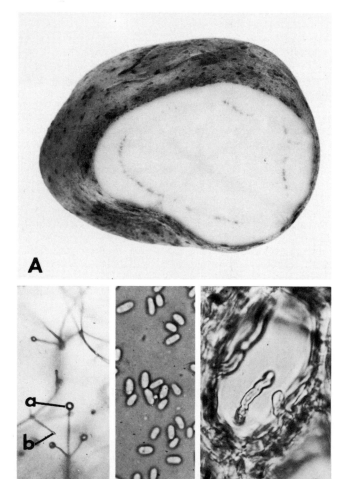

A

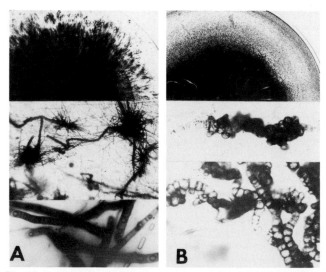

A   **B**

Fig. 68. **A,** Verticillium vascular discoloration in tuber. **B,** *Verticillium albo-atrum* in culture: **a,** water droplets surrounding conidia on tips of conidiophores; **b,** a single conidium (bar represents 50 μm). **C,** Conidia from culture; **D,** fungus mycelium in discolored xylem of tuber tissue (bar represents 10 μm).

Fig. 69. *Verticillium* isolates from potato. **A,** Dark mycelial type in one-month culture, at ×100 and ×375 approximately. **B,** Pseudosclerotial isolate in one-month culture, ×175 and ×200. (Courtesy D. B. Robinson et al 1957)

do not survive drying.

Both species are poor competitors and survive poorly in soil in the absence of suitable hosts. Infectious propagules of dark mycelium and pseudosclerotia are usually relatively short-lived, but survival time is influenced by soil type and is severely limited by anaerobic soil conditions. Infectious propagules germinate and produce conidia within hours.

Epidemiology

V. albo-atrum generally appers to be more pathogenic than *V. dahliae*. Warm soil temperatures (22–27° C) favor growth of *V. dahliae*. *V. albo-atrum* is relatively more pathogenic at lower temperatures; its range is 16–27° C. Crop rotation affects the buildup of soil inoculum, and three-year rotations have effectively reduced soilborne inoculum. Inoculum is reduced when cereals, grasses, and legumes are included in the rotation. When potatoes are grown several years in succession or in rotation with susceptible crops, inoculum increases in the soil. Other factors that favor incidence and severity of Verticillium wilt include planting highly susceptible cultivars such as Kennebec, discontinuation of the practice of burning potato tops, and a reduction in the formerly common practice of chemical treatment of seed potatoes.

Inoculum in field soil or in soil adhering to the surface of potato tubers is more important in initiating wilt symptoms than is inoculum from seed tubers with vascular discoloration. Paradoxically, seed tubers with vascular discoloration often produce plants as wilt-free and as vigorous as those from comparable seed tubers free from vascular discoloration.

Inoculum can be distributed long distances by contaminated soil adhering to seed surfaces and from field to field by contaminated equipment or irrigation water. Inoculum may also be airborne or spread from plant to plant by root contact.

Other Hosts

Isolations of *V. albo-atrum*, which may have also included *V. dahliae*, have been made from a very wide range of dicotyledonous plants, both woody and herbaceous. Many of these were symptomless. *V. dahliae* infects over 50 species of plants in 23 families. Grasses and other monocots are nonhosts of both species. Numerous common weeds are suscepts, including *Chenopodium album*, *Capsella bursa-pastoris*, *Taraxacum* spp., and *Equisetum arvense*.

Resistance

Varieties carrying different levels of resistance have been identified within *S. tuberosum* ssp. *tuberosum*. Certain cultivars react differently to varying conditions, including geographic location, abundance of inoculum, and type or strain of the fungus involved, suggesting that pathogenic strains of the causal agent exist. Even highest known levels of resistance break down with high inoculum density in the soil. *S. tuberosum* ssp. *andigena* is reported to have wilt resistance.

Control

1) Seed tubers contaminated with infested soil should be disinfested before being planted. Liquid seed treatments are more effective than dusts. Organic mercuries are very effective, but their use is generally prohibited.

2) Rotate potatoes with cereals, grasses, or legumes. Avoid rotation with highly susceptible solanaceous crops such as eggplant and most tomato cultivars.

3) Do not plant susceptible cultivars.

4) Control weed suscepts.

5) Systemic fungicides (benomyl or thiophanate-methyl) and nonsystemics (mancozeb, captan, or metiram) applied to seed tubers are reported effective.

6) Several soil treatments show promise; sodium methyldithiocarbamate, benomyl, and systemic insecticides (aldicarb, acephate) have delayed symptoms and increased some yields.

7) Several nematodes increase incidence and severity of Verticillium wilt. Soil fumigation with nematicides alone (tri-chloronitromethane or 1,3-dichloropropene and related compounds) or with chemicals that control both fungi and nematodes is effective.

Selected References

AYERS, G. W. 1974. Potato seed treatment for the control of verticillium wilt and Fusarium seed piece decay. Can. Plant Dis. Surv. 54:74–76.

BIEHN, W. L. 1970. Control of Verticillium wilt of potato by soil treatment with benomyl. Plant Dis. Rep. 54:171–173.

BUSCH, L. V. 1966. Susceptibility of potato varieties to Ontario isolates of *Verticillium albo-atrum*. Am. Potato J. 43:439–442.

EASTON, G. D., M. E. NAGLE, and D. L. BAILEY. 1975. Residual effect of soil fumigation with vine burning on control of Verticillium wilt of potato. Phytopathology 65:1419–1422.

ENGELHARD, A. W. 1957. Host index of *Verticillium albo-atrum* Reinke and Berth. (including *Verticillium dahliae* Kleb.). Plant Dis. Rep. Suppl. 244:23–49.

FRANK, J. A., R. E. WEBB, and D. R. WILSON. 1975. The effect of inoculum levels on field evaluations of potatoes for Verticillium wilt resistance. Phytopathology 65:225–228.

HIDE, G. A., and D. C. M. CORBETT. 1973. Controlling early death of potatoes caused by *Heterodera rostochiensis* and *Verticillium dahliae*. Ann. Appl. Biol. 75:461–462.

HOYMAN, W. G. 1974. Consequence of planting Norgold Russet seed infected with *Verticillium albo-atrum*. Am. Potato J. 51:22–25.

KHEW, K. L., and L. V. BUSCH. 1968. Soil temperature affects infection of potato and tomato by mixtures of DM and MS strains of *Verticillium albo-atrum*. Am. Potato J. 45:409–413.

KRAUSE, R. A., J. P. HUETHER, P. SATIROPOULOS, and L. E. ADAMS. 1975. Systemic insecticide control of aphids and potato Verticillium wilt. Plant Dis. Rep. 59:159–163.

MORSINK, F., and A. E. RICH. 1968. Interactions between *Verticillium albo-atrum* and *Pratylenchus penetrans* in the Verticillium wilt of potatoes. Phytopathology 58:401 (Abstr.).

POWELSON, R. L., and G. E. CARTER. 1973. Efficacy of soil fumigants for control of Verticillium wilt of potatoes. Am. Potato J. 50:162–167.

RICH, A. E. 1968. Potato diseases. Pages 397–437 in: O. Smith, ed. Potatoes: Production, Storing, Processing. Avi Publ. Co., Inc., Westport, CT.

ROBINSON, D. B., and G. W. AYERS. 1953. The control of Verticillium wilt of potatoes by seed treatment. Can. J. Agric. Sci. 33:147–152.

ROBINSON, D. B., R. H. LARSON, and J. C. WALKER. 1957. Verticillium wilt of potato in relation to symptoms, epidemiology, and variability of the pathogen. Wis. Agric. Exp. Stn. Res. Bull. 202. 49 pp.

SMITH, H. C. 1965. The morphology of *Verticillium albo-atrum, V. dahliae,* and *V. tricorpus*. N.Z. J. Agric. Res. 8:450–487.

THANASSOULOPOULOS, C. C., and W. J. HOOKER. 1970. Leaf and sprout infection of potato by *Verticillium albo-atrum*. Phytopathology 60:196–203.

WOOLLIAMS, G. E. 1966. Host range and symptomatology of *Verticillium dahliae* in economic, weed, and native plants in interior British Columbia. Can. J. Plant Sci. 46:661–669.

(Prepared by A. E. Rich)

Thecaphora Smut

Thecaphora smut is found in northern Central America (Mexico) and the northern regions of South America (Bolivia, Chile, Colombia, Ecuador, Peru, and Venezuela). Loss up to 80% is known.

Symptoms

Usually no aboveground symptoms are found. Affected tubers have warty swellings on the surface (Fig. 70A) and, when sectioned, reveal dark brown, locular sori pervading the interior (Fig. 71A and B, Plate 52). Galls resembling deformed tubers form directly from infections on the sprouts, stems (Fig. 70B), or stolons and, less frequently, on the tubers. Roots are not known to become infected.

Causal Organism

Angiosorus solani (Barrus) Thirum. & O'Brien (syn. *Thecaphora solani* Barrus) has locular sori, 1–1.2 mm in diameter, that are surrounded by a periderm six to eight cells deep and contain globose-ovoid spore balls $15–50 \times 12–40 \mu m$ in diameter (Fig. 71C). Each spore ball has two to eight rust-brown, subglobose to angular spores, $7.5–20 \times 8–18 \mu m$ that are easily separable into free spores when teased. Spore balls develop from sporiferous hyphae that form the locules and are pushed outward to fill the cavity.

Histopathology

Tumor growth, caused by hypertrophy of the outer phloem and parenchyma of the stem and stolon, consists largely of enlarged parenchymatous cells. The fungus is intercellular, 1.8 μm in diameter, producing clamp connections and thick branches. In the cambium, the fungus mycelium stimulates cell proliferation.

Disease Cycle

Details are unknown. Spore germination has not been observed. Smut is introduced by planting infected seed tubers; it may also be carried in irrigation water and infected soil.

Epidemiology

Smut is favored by high soil moisture and possibly high soil salinity. The planting of potatoes year after year increases disease when the pathogen is present.

Spores are believed to be long-lived in the soil. Although the disease was originally considered to be restricted to the Andes at elevations of 2,500–3,000 m, it is now known to persist and to be extremely serious in irrigated, sea-level desert terrain with high temperatures. Once introduced, the fungus would probably persist well in other potato-growing areas of the world.

Other Hosts

Infection develops on *S. tuberosum* ssp. *tuberosum* and ssp. *andigena*, *S. stoloniferum*, and *Datura stramonium*.

Control

1) The most effective control is to use cultivars with known resistance.

2) Planting smut-free seed is essential where the fungus is not present.

3) Long rotations reduce buildup of inoculum.

4) *Datura stramonium* becomes infected and should be eliminated in fields used for potatoes.

5) Removal of all smutted galls from an infested field reduces buildup of inoculum.

6) Strict quarantine, particularly of seed stock, should be exercised to avoid introduction into other potato-growing areas.

Selected References

BARRUS, M. F., and A. S. MULLER. 1943. An Andean disease of potato tubers. Phytopathology 33:1086–1089.

BAZAN De SEGURA, C. 1960. The gangrena disease of potato in Peru. Plant Dis. Rep. 44:257.

O'BRIEN, M. J., and M. J. THIRUMALACHAR. 1972. The identity of the potato smut. Sydowia Ann. Mycol. Ser. 2. 26:199–203.

UNTIVEROS, D. 1978. El carbon de la papa (*Thecaphora solani* Barrus), algunos aspectos de su sintomatologia y biologia del agente

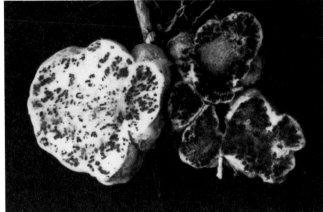

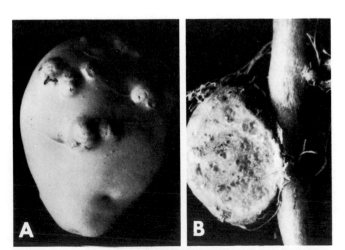

Fig. 70. Thecaphora smut: **A**, early infections on tuber; **B**, section through gall on stem. (Courtesy D. Untiveros)

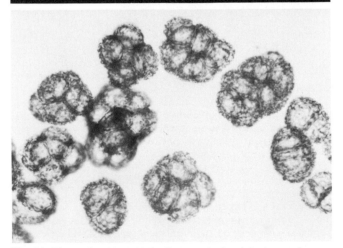

Fig. 71. *Thecaphora* smut infections: exterior, interior, and spore balls. (Courtesy R. Zachmann)

causal. Tesis Magister Scientiae en Fitopatologia. Universidad Nacional Agraria La Molina, Peru. 61 pp.

ZACHMANN, R., and D. BAUMANN. 1975. *Thecaphora solani* on potatoes in Peru: Present distribution and varietal resistance. Plant Dis. Rep. 59:928–931.

(Prepared by O. T. Page)

Common Rust

The disease occurs in restricted mountain valleys of the cool highlands of Mexico, Costa Rica, Venezuela, Colombia, Ecuador, and Peru, and possibly in Bolivia and Brazil. In Peru, it is usually on the eastern watershed of the Andes. Only in Ecuador is it of economic importance. The disease is most common at elevations of 3,000–4,300 m, although it occurs in a lower, warmer valley in Peru at an elevation of 2,700 m.

Symptoms

Round and occasionally elongate lesions usually develop on the underside of leaves as minute greenish white spots, 3–4 mm in diameter, although the longer axis of some oval lesions may reach 8 mm (Plate 53). Lesions later become cream with reddish centers, then tomato red, and finally rusty red to coffee brown (Fig. 72A). A chlorotic-necrotic halo may surround the lesions. Pustules protrude 1–3 mm and are matched by corresponding depressions on the upper side of the leaf. Defoliation results when hundreds of pustules form on a leaf.

Elongated and irregular lesions also occur on petioles and stems. Fruit and flowers are also affected.

Causal Organism

Puccinia pittieriana P. Henn., a short cycle (microcylic) rust, produces teliospores and sporidia (Fig. 72B). Sori are hypophyllous and gregarious. Teliospores are smooth, orange to brown, two-celled, broadly ellipsoid, slightly constricted at the septum, and $16–25 \times 20–35$ μm. The pedicel is 60×6 μm, and the hyaline sporidia are $8–18 \times 11–25$ μm.

Disease Cycle

Initial inoculum, probably from wild hosts, is windborne. In vitro, teliospores germinate in 1 hr to produce a promycelium, which, at temperatures above 15° C, usually continues to grow vegetatively. Below 15° C, most promycelia (basidia) give rise to four sporidia (basidiospores) in 3–24 hr. When detached, sporidia germinate immediately. First symptoms appear in 14–16 days on potato at temperatures of 16° C or below. Lesions are fully grown in 20–25 days. Teliospores mature 30–40 days after inoculation.

Epidemiology

Average temperatures around 10° C with 10–12 hr of free moisture on leaves are necessary for development and spread. The appearance of two distinct lesion sizes (2–4 mm and 4–8 mm in diameter) on a given variety and on different varieties suggests the existence of two races and two levels of susceptibility. Lesion size may vary with *Solanum* spp. Epidemic development may result in death of most plants and severe yield depression.

Other Hosts

Solanum demissum in Mexico is heavily infected. Tomato, the only other naturally infected host, is apparently more susceptible than potato. In greenhouse tests, *S. caripense* and *S. nigrum-americanum* are susceptible.

Control

Several carbamate fungicide sprays are effective when applied five times during the growing season at 14-day intervals at rates recommended for other foliage fungi.

(See Selected References following next section.)

Deforming Rust

This disease has been reported primarily from Peru, on the Pacific watershed range and at altitudes (2,378–3,172 m) overlapping those of common rust but generally somewhat lower.

Symptoms

Symptoms usually develop in mid to late growing season. Aecia initials first appear on leaves as smooth swellings that enlarge and rupture through the epidermis to form cups with an erose or lacerate peridial margin, which finally become saucer-shaped (Plate 54). The rim is yellowish and the center, a darker orange.

Rust pustules, consisting of crowded groups of aecia up to 100 mm across, are circular on the underside of leaf lamina and elongated along veins, petioles, and stems. They also affect flowers and fruit. The initial color, orange-red, turns at maturity to rusty brown. Lesions are largest on veins, petioles, and stems, causing pronounced enlargements and deformations such as thick curved leaves, stems swollen to double their size, and stems doubling over to the point of snapping. Defoliation and death of plants occasionally occur.

Causal Organism

Aecidium cantensis Arthur produces aecia, crowded in circular groups 5–10 mm across, on the underside of leaves.

Fig. 72. Common rust: **A,** mature red to brown colored sori; **B,** spores of *Puccinia pittieriana* (A, Courtesy C. R. French)

Each aecium is cupulate, 0.3–0.5 mm in diameter, with a colorless peridium. Aeciospores are angularly globoid or ellipsoid, 16–21 × 20–23 μm.

Epidemiology

The source of inoculum is probably cultivated potato, which is grown under irrigation in the drier season. Because this disease occurs below the elevation of severe frosts, the causal fungus may survive year around in plants and/or debris, but no research has been done on the overwintering of the pathogen.

Other Hosts

Potato is the only well verified host.

Control

Although Peruvian rust can be very damaging and recurs every rainy season in some localities, it is not widespread and thus not considered an important disease. No research on its control has been undertaken.

Selected References

ABBOTT, E. V. 1931. Further notes on plant diseases in Peru. Phytopathology 21:1061–1071.
ARTHUR, J. C. 1929. Las royas de los vegetales (Uredinales) del Perú. Est. Exp. Agric. Soc. Nac. Agrar. Lima, Peru. Bol. 2. 14 pp.
BURITICA, P., and J. ORJUELA. 1968. Estudios fisiológicos de *Puccinia pittieriana* Henn. causante de la roya de la papa (*Solanum tuberosum* L.) Fitotecnia Latinoamericana 5:81–88.
DIAZ, J. R., and J. ECHEVERRIA. 1963. Chemical control of *Puccinia pittieriana* on potatoes in Ecuador. Plant Dis. Rep. 47:800–801.
FRENCH, E. R., H. TORRES, T. A. de ICOCHEA, L. SALAZAR, C. FRIBOURG, E. N. FERNANDEZ, A. MARTIN, J. FRANCO, M. M. de SCURRAH, I. A. HERRERA, C. VISE, L. LAZO, and O. A. HIDALGO. 1972. Enfermedades de la Papa en el Perú. Bol. Tecn. 77. Est. Exp. Agric. La Molina. 36 pp.
HENNINGS, P. 1904. Einige neue pilze aus Costarica und Paraguay. Hedwigia 43:147–149.
THURSTON, H. D. 1973. Threatening plant diseases. Ann. Rev. Phytopathol. 11:27–52.

(Prepared by E. R. French)

Miscellaneous Diseases

Tuber Rots

In Germany and Italy, *Clonostachys araucariae* var. *rosea* tuber rot occasionally causes severe losses in storage following bad weather during harvest. Dark necrotic areas on tubers are surrounded by white mycelium with abundant conidia.

Armillaria dry rot is a minor problem in northern areas where potatoes are produced on recently cleared land. *Armillaria mellea* Vahl. ex Fr. causes hard brown, roughened areas somewhat corky in texture. Lesions are usually shallow, with dark brown to black rhizomorphs attached to such areas.

Xylaria tuber rot occurs in calcareous marl soils of Florida. Black rhizomorphs, threadlike to 3 mm thick, become firmly attached to the tuber surface. In the field, tuber invasion slowly progresses as more or less semicircular lesions. The problem, of minor importance, develops on recently cleared land.

Gilmaniella humicola causes minute brownish necrotic spots, 2–6 mm in diameter, around lenticels and eyes. Lesions remain shallow. Eyes are killed, making tubers unfit for seed.

Cylindrocarpon tonkinesis dry rot develops during the rainy season in India as brown patches on skin, later with white mycelium in or on affected tissue.

Heterosporium sp., usually found on potato leaves, also causes lesions on tubers.

Leaf Spots

Periconia sp., *Leptosphaerulina* sp., and *Didymella* sp. are simultaneously present in leaf spots, severely defoliating potatoes in Peru.

Chaetomium spp. leaf spot, superficially resembling early blight but without the targetlike markings, is reported in South Dakota. Necrosis and chlorosis develop around points of mechanical injury.

Selected References

COOK, A. A. 1954. A foliage disease of potato induced by *Chaetomium* species. Plant Dis. Rep. 38:403–404.
FABRICATORE, J. 1953. *Solanum tuberosum*, ospite casuale di un *Heterosporium*. Rev. Appl. Mycol. 1953:504.
JONES, W., and H. S. MacLEOD. 1937. Armillaria dry rot of potato tubers in British Columbia. Am. Potato J. 14:215–217.
KUHFUSS, K. H. 1957. *Clonostachys araucariae* Corda var. *rosea* Preuss an faulenden Kartoffelknollen. Nachrichtenbl. Pflanzenschutz. D. D. R. 11:144–146.
NYAK, M. L. 1964. A new organism causing dry rot of potatoes. Sci. Cult. 30:143–144.
PETTINARI, C. 1949. Azione patogena della *Clonostachys araucariae* Corda var. *rosea* Preuss su tuberi di *Solanum tuberosum*. Ann. Sper. Agrar. 3:665–672.
RUEHLE, G. D. 1941. A Xylaria tuber rot of potato. Phytopathology 31:936–939.
SAHAL, D. 1967. A new disease of potato tubers caused by *Gilmaniella humicola* Barron. Curr. Sci. 36:645–646.
VESSEY, J. C. 1975. Some potato diseases of the Peruvian humid tropics. Plant Dis. Rep. 59:1004–1007.

(Prepared by W. J. Hooker)

Mycorrhizal Fungi

The role of these obligate symbionts in potato growth and particularly their relation to tuberization has been extensively investigated. Only recently has their beneficial effect on potatoes been established. Inoculation in the root hair region with the endomycorrhizal fungus, *Glomus fasciculatus*, increased tuber yields and total plant weight.

Selected References

GRAHAM, S. O., N. E. GREEN, and J. W. HENDRIX. 1976. The influence of vesicular-arbuscular mycorrhizal fungi on growth and tuberization of potatoes. Mycologia 68:925–929.
PHILLIPS, J. M., and D. S. HAYMAN. 1970. Improved procedures for clearing roots and staining parasitic vesicular-arbuscular mycorrhizal fungi for rapid assessment of infection. Trans. Br. Mycol. Soc. 55:158–161.

(Prepared by W. J. Hooker)

Principles of Foliage Fungicide Application

In most parts of the world, sprays have superceded dusts in control of foliage diseases of potato. Dusts are easily and quickly applied without premixing with water, and dusters are usually cheaper, lighter, and simpler in construction and maintenance than sprayers. However, dusts drift easily, do not adhere well to foliage in the absence of moisture, and are not well suited to large-scale outdoor crop protection. On the other hand, sprays are less subject to drift, provide more uniform coverage, and stick better to foliage than dusts do, but accurate measurement and mixing of ingredients are essential to their effectiveness.

With few exceptions, the fungicides used to control foliage diseases of potato are protectant in their action. This means they must be applied to foliage before or at the same time as inoculum is deposited in order to prevent fungous spore germination, penetration, and subsequent disease development. Because protectant fungicides are not absorbed and translocated through the plant to any significant degree, they must be applied uniformly to as much of the foliage as possible. This implies uniformity both in horizontal distribution of fungicide across the spray swath and vertical distribution (penetration) through the plant canopy. Coverage of both upper and lower leaf surfaces is also essential. To achieve optimal coverage, attention should be paid to the following: use of the correct spray volume and pressure for which a sprayer has been designed; proper functioning of the sprayer as determined by cleanliness, wear of nozzle orifices, height of boom, accuracy of pressure gauge; spraying when the air is still; and not extending swath width beyond that specified for the sprayer.

Traditional hydraulic boom sprayers that apply large volumes (700–1,170 L/ha, 75–125 g/A) of dilute spray under high pressure (approximately 28 kg/cm^2, 400 psi) provide good horizontal and vertical distribution of fungicides. However, weight, high cost, large water requirement, and frequent need for filling have decreased their popularity in recent years. The present trend is towards low pressure (approximately 7.0–8.8 kg/cm^2, 100–125 psi), low volume (approximately 465 L/ha, 50 g/A or less) spraying, either with hydraulic or airblast (atomizing) machines. These can be effectively used if the manufacturer's recommendations are followed.

Spraying by aircraft has several advantages such as speed, saving of labor, ability to be used in fields too wet for ground equipment, absence of yield loss due to sprayer tracks, reduced spread of virus diseases, and low water requirements (usually 28–47 L/ha, 3–5 g/A). These advantages have increased acceptance of aircraft application particularly among large growers. But again, certain disadvantages must be considered, such as wind interference and risk of drift, unsuitability of small, irregularly shaped or topographically uneven fields, the hazard of physical obstacles, relatively poor vertical distribution of spray through the plant canopy, and sparse deposition of discreet particles of concentrated fungicide. This last makes it essential that fungicide deposits be redistributed, i.e., transferred from their original landing sites to unprotected leaf areas by moisture in the form of dew, rain, or irrigation. Although the addition of adjuvants to fungicide sprays has been promoted to improve redistribution, most fungicide formulations have adequate spreadability without being supplemented with such materials.

To provide as complete a protective blanket as possible, fungicide sprays must be applied at regular intervals to protect newly-expanded foliage and to supplement fungicide activity lost to dilution, photodegradation, oxidation, etc. Frequency of application depends on weather conditions, presence of disease in the vicinity, tenacity of the fungicide, varietal resistance, sprayer capability, and other factors. The usual interval employed by growers ranges from 5 to 14 days. In certain areas, disease forecasting systems are used to determine spray intervals.

Use of the correct amount of fungicide per unit area is vital to effective disease control. This requires careful attention to sprayer calibration and maintenance. Calibration means determining how much fungicide is applied by a sprayer to a particular area.

In summary, the four basic requirements for success are to apply the right chemical in the right amount at the right time in the right way to obtain maximum coverage.

Selected References

HORSFALL, J. G. 1956. Principles of Fungicidal Action. Chronica Botanica Co., Waltham, ME. 231 pp.
MARTIN, H. 1959. The Scientific Principles of Crop Protection, 4th ed. Edward Arnold, Ltd., London. 359 pp.
TORGESON, D. C., ed. 1967. Fungicides—An Advanced Treatise, Vol. 1. Academic Press, New York. 697 pp.

(Prepared by O. Schultz)

Tuber Seed Treatment

Chemical treatment of seed tubers before planting is neither a cure-all nor a replacement for the use of high quality seed, properly stored and handled. Rather, it is inexpensive insurance that partially protects seed from invasion by microorganisms present in the soil and on the tuber surface. Unfortunately, it does not kill microorganisms present within seed.

Possible benefits of seed treatment are: 1) control of storage diseases such as Fusarium dry rot, 2) control of seed piece decay when cut seed has to be held for an extended period before planting, 3) control of seed piece decay in the soil (usually caused by *Fusarium* spp.) when planting is done under adverse soil conditions (cool and wet) that impair suberization of the cut surfaces, and 4) partial control of diseases such as Rhizoctonia stem canker, scab, Verticillium wilt, and blackleg.

Prestorage treatment of seed, a relatively recent development, usually involves application of fungicides (e.g., thiabendazole) in the form of a fine mist as tubers enter the storage structure. Seed treatment after storage and before planting is most commonly done by dusting with rotating drum dusters, which generally provide better coverage than do the shaker types of treaters. Several ethylenebisdithiocarbamate fungicides are useful for dip treatment of seed tubers, but they are not commonly applied in this manner.

When whole seed is planted, or when cut seed is planted in a warm, moist seedbed promptly after cutting, whole tubers may be surface disinfested by dusting or dipping. This is only partially effective in controlling diseases such as Rhizoctonia stem canker, scab, and Verticillium wilt because their causal agents are commonly soilborne. Where cut seed is predominantly used, treating seed immediately after, rather than before, cutting is advisable (dusts are preferred). This protects the cut tuber surfaces from invasion by soilborne pathogens such as *Fusarium* spp. and provides partial surface disinfestation.

Cut seed, whether treated or not, should be planted immediately. If this is impossible, it should be stored in open containers such as potato crates stacked to provide ample ventilation. Burlap bags should be only half filled and also stacked for ventilation. Cut seed should be held at 10–16° C and high relative humidity (85–90%) for three to four days to allow wound cork to form over cut surfaces. If further storage is necessary, temperatures can be dropped to 5° C, but seed should be warmed again before planting to ensure vigorous sprouting. Insects should be excluded from cut seed, particularly adults of the seed corn maggot. Cut seed held for some time before planting may develop considerable seed piece decay and give poor stands.

Selected References

BOYD, A. E. W. 1975. Fungicides for potato tubers. Proc. 8th British Insecticide and Fungicide Conf. 3:1035–1044.
MOSHER, P. 1972. Treating seed potatoes for disease control. Spudlines, April 1972. Univ. of Maine. 3 pp.

(Prepared by O. Schultz)

Viruses

Virus diseases, although seldom lethal, reduce plant vigor and yield potential of seed tubers. Identification is complicated by wide symptom variation attributable to virus strain differences, plant response in relation to maturity and duration of infection, potato cultivar, and environmental influences. Differentiation of viruses may involve purification, electron microscopy, comparison of physical properties, electrophoresis, serology, and possibly other techniques. Recent developments in serology using the ELISA (enzyme-linked immunosorbent assay) technique provide an extremely sensitive test for virus presence and identity and have for the first time permitted serology with certain viruses difficult to detect in potato.

General References

BODE, O. 1958. Die Virosen der Kartoffel und des Tabaks. Pages 1–30 in: M. Klinkowski, ed. Pflanzliche Virologie, Vol. II. Akademie-Verlag, Berlin. 393 pp.

CLARK, M. F., and A. N. ADAMS. 1977. Characteristics of the microplate method of enzyme-linked immunosorbent assay for the detection of plant viruses. J. Gen. Virol. 34:475–483.

de BOKX, J. A., ed. 1972. Viruses of Potatoes and Seed-Potato Production. Pudoc, Wageningen, The Netherlands. 233 pp.

FERNANDEZ VALIELA, M. V. 1969. Introducción a la Fitopatología, Vol. I. Virus, 3rd ed. Pages 764-845. Inst. Nac. de Tecnol. Agropecuaria, Buenos Aires. 1,011 pp.

GIBBS, A., and B. HARRISON. 1976. Plant Virology: The Principles. John Wiley and Sons, New York. 292 pp.

KLINKOWSKI, M., and H. KEGLER. 1962. Viruskrankheiten der Kartoffel. Pages 1025-1138 in: R. Schick and M. Klinkowski, eds. Die Kartoffel, Vol. II. 2,112 pp.

MATTHEWS, R. E. F. 1970. Plant Virology. Academic Press, New York. 778 pp.

SMITH, K. M. 1972. A textbook of plant virus diseases, 3rd ed. Longman, London. 684 pp.

Potato Leafroll Virus

Potato leafroll, an aphid-transmitted disease is one of the most serious in potato and is responsible for high yield losses throughout the world wherever potatoes are grown. In Colombia and the Andes, the potato leafroll virus (PLRV) occurs in *Solanum andigena* potatoes; the disease has been known as "enanismo amarillo" for many years.

Symptoms

Primary symptoms (Plate 55) follow transmission by aphids. Symptoms appear mainly in the young leaves, which usually stand upright, roll, and turn slightly pale. In some cultivars, young leaves are pink to reddish starting at the margins. Rolling sometimes affects only the base of the leaflet rather than the whole leaflet. These symptoms may spread later to the lower leaves. Primary symptoms may be lacking with late season infection, making diagnosis in seed stocks a major problem.

Secondary symptoms (Plate 56) become evident when an infected tuber produces a plant. Lower leaflets are rolled and higher leaves are slightly pale. Leaves are stiff, dry, and leathery and make a crisp, somewhat paperlike sound when touched. Older leaves of some cultivars are pink and/or severely necrotic, especially at the margins. Plants are often noticeably stunted and rigid. Symptoms are less pronounced in the top of a plant with secondary infection than in one with primary infection, and secondary infection is more damaging to the plant. Severity of symptoms depends on the isolate of the virus, the potato cultivar, and the growing conditions.

In *S. tuberosum* ssp. *andigena* cultivars, secondary symptoms (enanismo amarillo) tend to be somewhat different from those in *S. tuberosum* ssp. *tuberosum*, consisting of a marked upright habit of growth, stunting, and a marginal and interveinal chlorosis of leaflets (Plate 57), especially of dwarfed upper leaves. Rolling of lower leaves is often lacking. Secondary symptoms in hybrids of *S. tuberosum* ssp. *andigena* × *S. tuberosum* ssp. *tuberosum* consist of the rolling of lower leaves typical of *S. tuberosum* ssp. *tuberosum* cultivars. In some cultivars, however, distinct interveinal and marginal chlorosis occurs as in *S. tuberosum* ssp. *andigena* cultivars.

Internal net necrosis (Figs. 28B and 73A), visible to the unaided eye when the tuber is cut, is particularly marked in certain American cultivars such as Russet Burbank, Green Mountain, and Norgold Russet. This net necrosis is found in tubers from plants with primary, secondary, or tertiary infection. The color may vary from light translucent to dark. A definite trend toward a less severe type of infection from primary to tertiary has been found in tubers of Russet Burbank.

Causal Agent

PLRV has icosahedral particles 24 nm in diameter (Figs. 73D and E).

Transmission is successful only by aphids and grafting; artificial inoculation by mechanical means has never been accomplished. The virus is transmitted by aphids in a persistent (circulative) manner (i.e., after an aphid has fed on an infected plant it normally transmits the virus for the rest of its life).

Isolates have not been differentiated by serological, physical, or chemical properties nor by vector specificity or transmission efficiency. No differential plants are known. All strains give the same type of symptoms. Isolates vary in severity of symptoms on potato cultivars and on *Physalis floridana*, but existence of strains has not been well defined. Some workers refer to three strains called severe, moderate, and mild; others refer to four strains called No. 1, 2, 3, and 4. A fifth isolate, differing in symptom severity from No. 1 on *P. floridana* but not on other hosts, has been isolated.

Histopathology

From the moment of symptom appearance, infections are always accompanied by phloem necrosis, which consists of thickening of the walls of the primary phloem cells in the stem and petioles (Fig. 73B). Accumulation of callose is often pronounced around the sieve plates in the phloem of tubers and stems (Fig. 73C). Stiffness of leaves in primary and secondary infections is a consequence of starch accumulation in leaf cells.

Epidemiology

The virus is tuberborne and is also efficiently transmitted in a persistent manner by aphids that colonize potatoes, *Myzus persicae* being the most efficient. Infection efficiency increases with the length of feeding. The virus is spread over long distances by windborne winged aphids and over short distances by nonwinged aphids moving from plant to plant. Aphid transmission occurs from tuber to tuber during seed storage, especially in tropical countries. Plants from infected tubers and diseased volunteer potato plants also serve as virus sources. Moderate temperatures and dry weather favor spread. Plants become resistant to infection with age, and occasionally some tubers of plants infected late in the season escape infection.

Other Hosts

Several plant species, mostly in the Solanaceae, are known as hosts.

P. floridana is a suitable indicator, test, and propagation host that reacts with interveinal chlorosis, darkening of the veinal areas, and a slight cupping at the first two or three true leaves. Infections at a later stage cause plants to become somewhat pale.

Datura stramonium is favored by some workers, especially in the more tropical regions, as a test and propagation host in which chlorosis develops. In Brazil, alternative hosts to PLRV have been demonstrated, including tomato, *D. stramonium*, and a *Physalis* species. *D. stramonium* also seems to be a reservoir for the virus in the Andean region, where it is a common weed. No evidence indicates that hosts other than potato act as a virus reservoir in the temperate regions.

Nonsolanaceous hosts are found in the Amaranthaceae.

Tuber Indexing

Symptom development in eye sprouts can be used to index for virus infection in postharvest glasshouse tests. The often-used Igel-Lange test (deep blue staining of phloem sieve tubes after 10 min in 1% aqueous resorcin blue) is based on increased deposits of callose in the phloem (Fig. 73C). Neither method is completely reliable because symptoms are not always produced and because callose, which is not always observed in infected tubers, may also be found in healthy tubers.

Electron microscope techniques alone are unreliable because of the low concentration of virus particles in infected plants. In combination with serological techniques, electron microscopy is efficient.

The recent application of ELISA to detect the virus in tuber and other plant material has resulted in improved indexing techniques. A reliable method, although laborious to perform, involves aphid transmission tests on eye sprouts, using indicator plants such as *P. floridana*.

A technique in which leaves, stems, and small pieces of tuber are grafted to *D. stramonium* is now used on a large scale in seed multiplication work in Brazil. This method is considered to be less time-consuming than that using aphids and *P. floridana*.

Control

1) Breeding resistant cultivars has met only limited success. Resistance is determined by many genes with additive effects and can only gradually be built up. Varying levels of resistance are found in some seedlings. Crossing of unrelated resistant parents may result in a higher proportion of resistant offspring than does inbreeding. Resistance is also found in progenies from *S. demissum, S. andigena, S. acaule, S. chacoense, S. stoloniferum,* and *S. etuberosum*.

2) Disease-free seed tubers are essential for maximum production. Seed tubers with a low percentage of infection can be produced in areas where sources of virus are limited and aphids appear late in the season. Seed plots should be harvested as early as possible, compatible with reasonable yields, to avoid late season aphid transmission. Dates for lifting may be determined by the number of aphids caught in yellow traps or assessed by other methods.

3) To minimize infection, measures such as clonal selection, early planting of virus-free tubers (from seed certification programs), and early lifting and roguing of infected plants and killing or removal of volunteer plants in and around the field can be used.

4) Aphids should be controlled by toxic sprays or systemic insecticides. Application of granulated systemic insecticides has been successful in certain cases. Spread cannot be controlled by oil sprays.

5) Tubers may be freed of the virus by heating, e.g., 25 days at 37.5°C.

Selected References

BACON, O. G., V. E. BURTON, D. L. McLEAN, R. H. JAMES, W. D. RILEY, K. G. BAGHOTT, and M. G. KINSEY. 1976. Control of the green peach aphid and its effect on the incidence of potato leaf roll virus. J. Econ. Entomol. 69:410–414.

CHIKO, A. W., and J. W. GUTHRIE. 1969. An hypothesis for selection of strains of potato leafroll virus by passage through *Physalis floridana*. Am. Potato J. 46:155–167.

CUPERTINO, F. P., and A. S. COSTA. 1967. Determinação do vírus do enrolamento em hastes velhas de batatal para sementes. Bragantia 26:181–186.

DAVIDSON, T. M. W. 1973. Assessing resistance to leafroll in potato seedlings. Potato Res. 16:99–108.

de BOKX, J. A. 1967. The callose test for the detection of leafroll virus

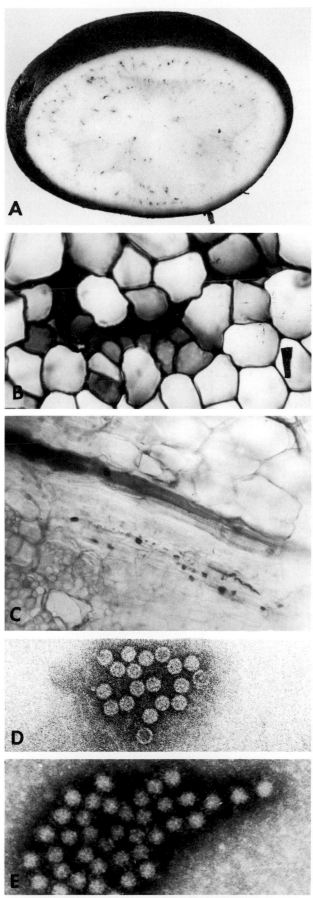

Fig. 73. Potato leafroll virus (PLRV): **A,** phloem net necrosis of tuber; **B,** phloem necrosis in petiole (bar represents 10 μm); **C,** callose at stolon end of infected tuber (Igel-Lange stain); **D,** PLRV particles, 24 nm in diameter, negatively stained with phosphotungstic acid and (**E**) precipitated with PLRV antiserum and similarly stained. (C–E, Courtesy D. Peters)

in potato tubers. Eur. Potato J. 10:221–234.

DOUGLAS, D. R., and J. J. PAVEK. 1972. Net necrosis of potato tubers associated with primary, secondary, and tertiary infection of leafroll. Am. Potato J. 49:330–333.

MAAT, D. Z., and J. A. de BOKX. 1978. Potato leafroll virus: Antiserum preparation and detection in potato leaves and sprouts with the enzyme-linked immunosorbent assay (ELISA). Neth. J. Plant Pathol. 84:149–156.

PETERS, D. 1970. Potato leafroll virus. No. 36 in: Descriptions of Plant Viruses. Commonw. Mycol. Inst., Assoc. Appl. Biol., Kew, Surrey, England.

SMITH, K. M. 1972. A Textbook of Plant Virus Diseases, 3rd ed. Longman: London. 684 pp.

(Prepared by D. Peters and R. A. C. Jones)

Potato Virus Y

Potato virus Y (PVY) causes rugose mosaic (Fig. 74A). Common strains of the virus (PVY^O) are found worldwide, and necrotic strains (PVY^N) occur in Europe, including the USSR, and in parts of Africa and South America. Strains belonging to the type virus C (PVY^C) probably occur in Australia and some parts of Europe, although extensive reports are not available.

Symptoms

Symptoms in potato vary widely with virus strain and potato cultivar, ranging in severity from weak symptoms to severe foliage necrosis to death of infected plants. In general, PVY^O and PVY^C cause much more severe symptoms than does PVY^N. PVY^N induces vague mottle in plants with current-season

Fig. 74. Rugose mosaic: **A**, in cultivar Bintje; **B**, tuber symptoms of unnamed seedling with potato virus Y^O. (Courtesy J. A. de Bokx)

(primary) infection and in those from infected tubers (secondary infection). If infection occurs late in the season, foliage symptoms may not appear, but tubers from such plants may carry the disease.

Primary symptoms of PVY^O, depending on potato cultivar, are necrosis, mottling, or yellowing of leaflets, leaf dropping, and sometimes premature death. Necrosis, which starts as spots or rings on the leaflets (Plate 58), may cause leaves to collapse and either drop from the plants (leafdrop streak) or remain clinging to the stem, resembling palm trees. Sometimes these symptoms appear only on a single shoot in a hill.

Plants with secondary PVY^O infection are dwarfed, and leaves are mottled and crinkled. Sometimes foliage and stem necrosis occurs. Necrosis is usually more severe after primary than after secondary infection. The foliage symptom is a mosaic (Plate 59), which differs from that induced by potato virus A (mild mosaic) in that discolored areas are smaller and more numerous. Leaf mottling may be masked at very low (10°C) and high (25° C) temperatures, but at high temperatures the disease can be identified by the crinkling and rugosity of the foliage (Plate 60).

PVY^C evokes stipple-streak symptoms in several cultivars. Affected plants are dwarfed and may die prematurely.

A correlation generally exists between symptoms in the foliage and those in tubers. Weak mosaic symptoms in the foliage, as commonly induced by PVY^N strains, are not accompanied by symptoms in the tubers. Cultivars that react with necrosis in the foliage upon infection to PVY^O sometimes show light brown rings on the skin of tubers (Fig. 74B). PVY^C strains may induce internal and external necrosis in some cultivars.

Causal Agent

PVY is a virus with flexuous, helically constructed particles, 730 × 11 nm (Fig. 75A).

Many groups of strains can be distinguished according to severity of systemic symptoms in tobacco, *Physalis floridana*, potato, and other hosts. The main groups are called PVY^O, PVY^N, and PVY^C. Although aphids can transmit PVY^C, they transmit some strains only with great difficulty, if at all, and others as readily as they transmit PVY^N.

Histopathology

Pinwheel inclusions observed in electron microscopy occur in tissue of tobacco and potato systemically infected with PVY (Fig. 75B).

Epidemiology

Spread of PVY depends mainly on the presence of winged aphids. The virus is borne on the stylet and is transmitted within a few seconds in a nonpersistent way by many aphid species. at least 25 are mentioned as capable of transmitting PVY, but little is known about their efficiency. *Myzus persicae* is the most efficient species in many areas and seasons.

PVY is considered one of the most damaging potato viruses in causing yield depression. PVY^O and PVY^C may cause complete failure of a potato crop. In combination with potato virus X, PVY is generally even more destructive, producing the rugose mosaic disease.

Other Hosts

Many plant species, mostly in the Solanaceae, but also in the Chenopodiaceae and Leguminosae are hosts of PVY. Perennials seldom act as virus reservoirs in nature. In potato-growing areas, the potato itself, as volunteer plants, may be considered a reservoir host.

In *Nicotiana tabacum* (tobacco), most strains produce vein clearing followed by mottling. The necrotic PVY^N strain induces mottling and bronzing of the midribs and necrosis of leaf ribs.

In *P. floridana*, strains of PVY^C and PVY^O cause local and systemic necrosis in young plants. Rapid death follows inoculation with PVY^C, whereas PVY^N induces mosiac symptoms only.

Potato cultivars, e.g., Duke of York, may be useful for further identification because after being grafted with PVY^N, PVY^O, and PVY^C, they react with mottling, rugose mosaic, and stipple streak, respectively. No hosts are known to separate the different strains.

Solanum demissum 'Y' of Cockerham or 'A6' (*S. demissum* × *S. tuberosum* 'Aquila') react to PVY with local lesions. Detached leaves are often used in routine tests.

Datura stramonium or *D. tatula* are resistant to PVY and serve for elimination of PVY in certain virus complexes.

Seed Propagation

Mature plant resistance may be of importance in seed-potato production procedures. When infection by aphids takes place in an advanced stage of plant growth, the virus may not spread to tubers; frequently only a few tubers per plant become infected. Mature plant resistance is much less effective against PVY^N

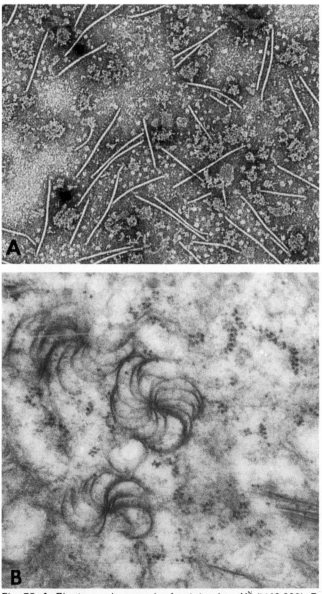

Fig. 75. **A,** Electron micrograph of potato virus Y^N (×40,000); **B,** pinwheel inclusion bodies demonstrated by thin section, electron microscopy (×72,000), in tissues infected with either potato virus Y or A. (A, Courtesy H. Huttinga; B, courtesy J. A. de Bokx)

than against PVY^O strains.

Breeding and growing cultivars resistant to PVY is the best way of combating the disease. Extreme resistance, which protects plants from all strains of the virus, occurs in *S. chacoense* and *S. stoloniferum*. Seedlings with good commercial qualities have been derived from these.

In areas where virus sources are limited and vectors are present only in certain periods of the growing season, seed with a tolerable percentage of infection can be produced. Certification schemes use combinations of measures such as clonal selection, planting virus-free seed at an early date, roguing, and early lifting. This may be followed by tuber indexing. Laboratory testing, such as leaflet inoculation of 'A6' and serology, can be included.

Heat treatment and meristem culture may be used for freeing virus-infected clones.

Preventing spread of PVY strains by chemical control of aphid vectors, including the use of systemic insecticides, is not possible because aphids have short periods of PVY acquisition and infection feeding.

Control

1) Plant clean seed free from tuberborne infection.

2) Use resistant cultivars.

3) Plant early, and rogue diseased plants.

4) Prevent heavy aphid populations in the field by applying insecticides as foliage sprays and systemic soil treatments. Application of oil spray to the foliage has been useful experimentally.

5) Time harvesting operations to precede heavy aphid flights as determined by yellow pan traps.

Selected References

COCKERHAM, G. 1957. Experimental breeding in relation to virus resistance. Pages 199–203 in: F. Quak, J. Dijkstra, A. B. R. Beemster, and J. P. H. Van der Want, eds. 1958. Proc. Third Conf. Potato Virus Dis., 24–28 June, 1957. H. Veenman and Zonen, Lisse-Wageningen, The Netherlands. 282 pp.

de BOKX, J. A., ed. 1972. Viruses of Potatoes and Seed-Potato Production. Pudoc, Wageningen, The Netherlands. 233 pp.

de BOKX, J. A., and P. G. M. PIRON. 1978. Transmission of potato virus Y^C by aphids. Eur. Assoc. Potato Res. 7th Triennial Conf. pp. 244–245 (Abstr.).

DELGADO-SANCHEZ, S., and R. G. GROGAN. 1970. Potato Virus Y. No. 37 in: Descriptions of Plant Viruses. Commonw. Mycol. Inst., Assoc. Appl. Biol., Kew, Surrey, England.

EDWARDSON, J. R. 1974. Some properties of the potato virus Y-group. Fl. Agric. Exp. Stn. Monograph Ser. No. 4. 398 pp.

EDWARDSON, J. R. 1974. Host ranges of viruses in the PVY-group. Fl. Agric. Exp. Stn. Monograph Ser. No. 5. 225 pp.

WEBB, R. E., and D. R. WILSON. 1978. *Solanum demissum* P. I. 230579, a true seed diagnostic host for potato virus Y. Am. Potato J. 55:15–23.

(Prepared by J. A. de Bokx)

Potato Virus A

Potato virus A (PVA), which causes mild mosaic, is widespread in most potato-growing areas. PVA may decrease yield of infected potatoes by up to 40%.

Symptoms

The mild mosaic symptom induced in leaves of many potato cultivars is a chlorotic mottle, sometimes severe, in which yellowish or light-colored irregular areas alternate with similar areas of darker than normal green (Fig. 76A). Mottled areas vary in size and lie both on and between the veins. A slight rugosity of the leaf surface can usually be observed, and the margins of the leaflets may become wavy. Infected leaves as a whole look shiny. Severity of symptom expression depends largely on weather conditions, the potato cultivar, and the strain of virus A.

Infected plants usually seem open because stems bend outward. At high temperatures and in bright sunlight, symptoms are more difficult to recognize than in cloudy cool weather and may be completely masked. Normally, infected plants produce symptomless tubers.

Causal Agent

PVA, a member of the potato virus Y (PVY) group, has particles as flexuous filaments, 730×15 nm. PVA is serologically related to PVY, although the degree of relationship is difficult to assess because reciprocal heterologous tests do not give the same results.

According to symptoms in potato and *Nicandra physalodes*, strains of PVA may be classed into three groups: mild, moderately severe, and severe.

Histopathology

Pinwheel inclusions occur in tissue of tobacco and potato

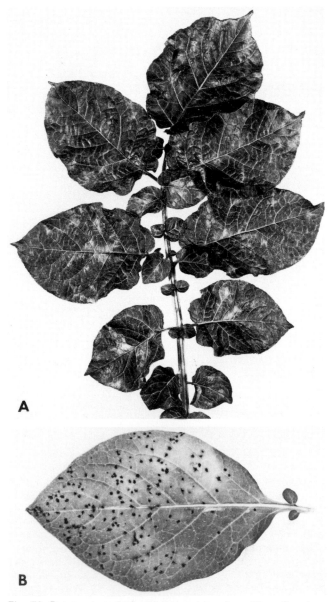

A

B

Fig. 76. Potato virus A: **A,** mild symptoms in cultivar Saucisse Rouge; **B,** necrotic local lesions in A6, a *Solanum demissum* × *S. tuberosum* hybrid. (Courtesy J. A. de Bokx)

systemically infected with PVA (Fig. 75B).

Epidemiology

PVA is transmissible by at least seven species of aphids (including *Aphis frangulae, Macrosiphum euphorbiae,* and *Myzus persicae*) in the nonpersistent manner, by grafting, and by inoculation with sap. The latter, however, usually yields poor results because of instability of the virus. Therefore, transmission by foliage contact in the field probably does not take place. Many cultivars, e.g., Bintje, Katahdin, Kennebec, and Sebago, are field resistant (hypersensitive) to all known strains of PVA. For this reason, PVA is considered less harmful than PVY. Plants extremely resistant to PVY, obtained from *S. chacoense* or *S. stoloniferum,* are also extremely resistant to PVA.

Testing for the presence of PVA by using 'A6' or meristem culture for freeing virus-infected stock is useful in seed improvement programs. PVA cannot be detected by present serological methods; however, for serological identification of PVA, the ELISA technique can be applied.

Other Hosts

Hosts of PVA are limited to the Solanaceae.

In *Nicotiana tabacum* cv. Samsun, virus strains produce vein clearing and diffuse mottle, whereas in *N. tabacum* cv. White Burley, vein clearing and vein banding are induced.

'A6' (*S. demissum* × *S. tuberosum* cv. Aquila) develops local lesions (Fig. 76B), which may be distinguished from those of PVY in detached leaf culture under controlled environment.

S. demissum 'A' of Cockerham shows local lesions only after inoculation with PVA.

Control

1) Plant clean seed free from tuberborne infection.
2) Use resistant cultivars.
3) Plant early, and rogue diseased plants.
4) Prevent heavy aphid populations in the field by applying insecticides as foliage sprays and systemic soil treatments. Application of oil spray to the foliage has been useful experimentally.
5) Time harvesting operations to precede heavy aphid flights as determined by yellow pan traps.

Selected References

BARTERLS, R. 1971. Potato virus A. No. 54 in: Descriptions of Plant Viruses. Commonw. Mycol. Inst., Assoc. Appl. Biol., Kew, Surrey, England.

COCKERHAM, G. 1957. Experimental breeding in relation to virus resistance. Pages 199–203 in: F. Quak, J. Dijkstra, A. B. R. Beemster, and J. P. H. Van der Want, eds. Proc. Third Conf. Potato Virus Dis., 24–28 June, 1957. H. Veenman and Zonen, Lisse-Wageningen, The Netherlands. 282 pp.

de BOKX, J. A., ed. 1972. Viruses of Potatoes and Seed-Potato Production. Pudoc, Wageningen, The Netherlands. 233 pp.

FERNANDEZ VALIELA, M. V. 1969. Introducción a la Fitopatologiá, Vol. I. Virus, 3rd ed. Pages 764–845. Inst. Nac. Tecnol. Agropecuaria, Buenos Aires. 1,011 pp.

MAAT, D. A., and J. A. de BOKX. 1978. Enzyme-linked immunosorbent assay (ELISA) for the detection of potato viruses A and Y in potato leaves and sprouts. Neth. J. Plant Pathol. 84:167–173.

WEBB, R. E., and R. W. BUCK, Jr. 1955. A diagnostic host for potato virus A. Am. Potato J. 32:248–252.

(Prepared by J. A. de Bokx)

Potato Virus X

Potato virus X (PVX) occurs wherever potatoes are grown. It is the most widespread of potato viruses and often completely infects certain commercial stocks, with yield reductions estimated to range up to more than 15%.

Symptoms

PVX may be latent, without foliage symptoms or apparent effect on plant vigor except when closely compared to PVX-free stocks, or it may show mild mottle (Plate 61) to severe or rugose mosaic (Fig. 77A), with dwarfing of the plant and reduced leaflet size. Certain combinations of PVX strain and host genotype cause extensive top necrosis (Fig. 77B), which may kill part or all of the plant and cause tuber necrosis. In combination with potato viruses A or Y, crinkling, rugosity, or necrosis may develop.

Causal Agent

PVX particles are flexuous filaments, 515 × 13 nm, with helical (pitch, 3.4 nm) substructure (Fig. 78D). Single-stranded ribonucleic acid, with molecular weight of 2.1×10^6, comprises 6% of the particle weight. Thermal inactivation is 68–76, depending on the strain; dilution end point is 10^{-5}–10^{-6}; longevity in vitro is several weeks.

PVX is strongly immunogenic and is sufficiently homogeneous to permit several serological methods to be used in identification of naturally infected plants.

PVX isolates have been grouped into strains by cross absorption serology (four groups), temperature of inactivation (three groups), and necrotic reaction within *S. tuberosum* spp. *tuberosum* (four groups). Strains related by any of these groupings may vary considerably in the intensity of the symptoms they cause.

Within potato leaf cells, PVX forms large amorphous inclusions readily observable with light microscopy. These inclusions, examined in thin section by electron microscopy, contain virus particles interspersed between alternating layers of curved or rolled laminate inclusion components.

Epidemiology

PVX is, with few exceptions, transmitted by tubers in susceptible cultivars. Transmission through sap inoculation is readily accomplished by contact of plant parts in the field due to wind, animals, or machinery; by root contact; by sprout to sprout contact; by the cutting knife before planting; and by biting insects (grasshoppers). Zoospores of *Synchytrium endobioticum* are reported to transmit the virus. Aphids are not known to transmit PVX, nor is the virus known to be transmitted through true seed.

Symptoms in most plants are enhanced by low temperatures of 16–20°C and are mild or may be masked at temperatures above 28°C.

Cultivars may be freed from PVX by meristem culture of sprout tips grown at 32–36°C.

Other Hosts

PVX isolates cause widely different symptom severity ranging from mild to severe. Mild isolates may spontaneously become severe or vice versa. PVX is chiefly systemic in the Solanaceae, usually produces local lesions in the Chenopodiaceae or Amaranthaceae, and also infects certain Leguminosae.

In *Nicotiana tabacum* (White Burley or Samsun types), infection is systemic, with mottles or ring spots (Fig. 78B and C, Plate 62). The plant is useful for virus propagation.

Datura stramonium and *D. tatula* show systemic infection, latent to severe mottle, and some leaf necrosis.

Gomphrena globosa is a local lesion host (Fig. 78A) useful for quantitative infectivity assay.

Resistance

Comprehensive or extreme resistance, earlier called "immunity," is determined by a single dominant gene (Rx). This gene—or one similar—confers resistance to all strains of PVX except one recently isolated in the Andes. It is present in selections of *Solanum acaule* and its derivatives, including the cultivar Saphir, and in certain selections of *S. tuberosum* ssp. *undigena*, which include C. P. C. 1673, the cultivar Villaroela, and its probable derivative S.41956.

Field immunity in *S. tuberosum* ssp. *tuberosum*, determined by dominant genes, is a strong necrotic, often top necrotic, reaction to infection. The Nx:nb genetic type confers resistance to Cockerham's (1954) PVX groups 1 and 3, nx:Nb to groups 1 and 2, and Nx:Nb to groups 1–3. Genes conferring similar resistance are present in certain other tuber-bearing *Solanum* spp.

Certain cultivars, although susceptible, apparently prevent completely free PVX movement within the plant, permitting certain plant parts to escape infection. By contrast, other cultivars are so thoroughly invaded that obtaining virus-free parts for propagation is difficult.

A strain of PVX (X_{HB}) recently isolated in the Andean highlands differs markedly from previously described PVX strains by infecting 1) plants with the Rx gene (USDA S.41956, Saco, Saphir, and resistant selections of *S. acaule*) and 2) *G. globosa* without symptoms in inoculated leaves. This strain does not become systemic in *G. globosa*.

Fig. 77. Potato virus X: **A,** latent strain in branch at right and severe rugose strain at left; **B,** top necrosis in field-resistant Epicure following graft inoculation with potato virus X. (A, Courtesy J. Munro)

Fig. 78. Potato virus X infections: **A,** *Gomphrena globosa;* **B,** Samsun tobacco; **C,** Havana tobacco inoculated leaf with clear ring spots and leaf of same plant with systemic vein clearing. **D,** Electron micrograph of potato virus X particles (×83,000). (B, Courtesy J. Munro)

Control

1) Use PVX-free seed and avoid contamination through contact with infected plants or tubers.

2) Use resistant cultivars where possible. Tolerant cultivars are useful. Many high yielding seed stocks, completely infected with PVX, produce plants with excellent vine type and high tuber quality.

Selected References

BERCKS, R. 1970. Potato Virus X. No. 4 in: Descriptions of Plant Viruses. Commonw. Mycol. Inst., Assoc. Appl. Biol., Kew, Surrey, England.

COCKERHAM, G. 1954. XIII. Strains of potato virus X. Pages 82–92 in: E. Streutgers, A. B. R. Beemster, D. Noordam, and J. P. H. Van der Want, eds. 1955. Proc. Second Conf. Potato Virus Dis., 25–29 June, 1954. H. Veenman and Zonen, Lisse-Wageningen, The Netherlands. 193 pp.

COCKERHAM, G. 1970. Genetical studies on resistance to potato viruses X and Y. Heredity 25:309–348.

DELHEY, R. 1974. Zur Natur der extremen Virusresistenz bei der Kartoffel. Phytopathol. Z. 80:97–119.

MELLOR, F. C., and R. STACE-SMITH. 1970. Virus strain differences in eradication of viruses X and S. Phytopathology 60:1587–1590.

MOREIRA, A., R. A. C. JONES, and C. E. FRIBOURG. 1980. Properties of a resistance-breaking strain of potato virus X. Ann. Appl. Biol. 95:93–103.

MUNRO, J. 1954. Maintenance of virus X-free potatoes. Am. Potato J. 31:73–82.

MUNRO, J. 1961. The importance of potato virus X. Am. Potato J. 38:440–447.

SHALLA, T. A., and J. F. SHEPARD. 1972. The structure and antigenic analysis of amorphous inclusion bodies induced by potato virus X. Virology 49:654–657.

SHEPARD, J. F., and G. A. SECOR. 1969. Detection of potato virus X in infected plant tissue by radial and double-diffusion tests in agar. Phytopathology 59:1838–1844.

WRIGHT, N. S. 1970. Combined effects of potato viruses X and S on yield of Netted Gem and White Rose potatoes. Am. Potato J. 47:475–478.

(Prepared by J. Munro)

Potato Virus M

Potato virus M (PVM) is found worldwide in potato cultivars. It is more important in Eastern Europe and the USSR than in other parts of the world. Its economic significance in North America is uncertain.

The disease was first named potato leafrolling mosaic. Early descriptions were probably not made on PVM alone because the virus frequently occurs with potato viruses X and/or S. The identity of PVM has only recently been clearly established. PVM causes the latent paracrinkle disease in the King Edward cultivar.

Symptoms

Aboveground symptoms range from very slight to severe (Fig. 79 A–C, Plate 63) and include mottle, mosaic, crinkling, and rolling of leaves; stunting of shoots; leaflet deformation and twisting; and some rolling of the top of the plant. Severity is influenced by virus strain, potato cultivar, and environmental conditions. Necrosis of petioles and stems may develop in certain potato cultivars.

Causal Agent

PVM particles are straight to slightly flexuous rods 650×12 nm. The thermal inactivation point is between 65 and 70° C; dilution end point is 10^{-2}–10^{-4}; and infectivity is retained at 20° C for two to four days. The virus is a good immunogen.

Cytopathology and Histopathology

Numerous PVM rods and virus aggregates occur in the

cytoplasm of infected potato cells (Fig. 79E). These include neither pinwheels nor pinwheel-related structures. No virus particles and/or aggregates are found in chloroplasts, mitochondria, or nuclei.

Epidemiology

Transmission by mechanical inoculation with infective sap or by tuber or stem grafting can be achieved with ease. True seed transmission has not been demonstrated. Most PVM strains are aphid-transmissible in the nonpersistent manner by *Myzus persicae* and less efficiently by *Macrosiphum euphorbiae*, *Aphis frangulae*, and *A. nasturtii*. Strains differ in efficiency of transmission by aphids. Symptoms are masked at temperatures or approximately 24° C and above.

Other Hosts

PVM infects mainly the Solanaceae but also members of the Chenopodiaceae and Leguminosae.

Local dark green or yellow spots in *Chenopodium quinoa* and local chlorotic rings or necrotic spots in *Gomphrena globosa* are produced by some PVM isolates only. Symptoms in *Datura metel* are local chlorosis or necrosis and later, systemic necrosis. *Lycopersicon chilense* exhibits epinasty, distortion, stunting, and abscission. *L. esculentum* (tomato) is susceptible to PVM but remains symptomless and is immune to potato virus S.

In *Solanum rostratum*, necrosis is systemic. In *Nicotiana debneyi*, local brown ringlike necrotic lesions are produced by some PVM isolates only. *Phaseolus vulgaris* cv. Red Kidney exhibits local necrotic lesions on the primary leaves (Fig. 79D). French bean is a convenient and reliable local lesion host for quantitative investigations of PVM.

In *Vigna sinensis*, local brown necrotic lesions are produced. *N. tabacum* and *Physalis floridana* are nonsusceptible.

To distinguish between PVM strains, *L. chilense* and the potato cultivars Kennebec and Prinslander are useful.

Control

1) Use disease-free virus-tested seed tubers.
2) Control aphid populations.
3) Rogue infected plants when first found in the field.
4) Infected potato can be freed by apical meristem culture only or by combining heat treatment with axillary bud culture.

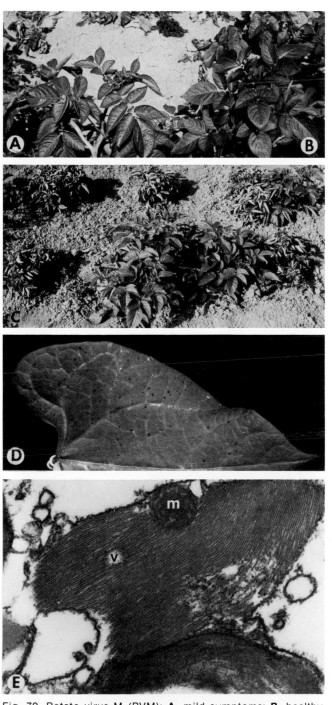

Fig. 79. Potato virus M (PVM): **A**, mild symptoms; **B**, healthy control; **C**, severe symptoms in Kennebec; **D**, local necrotic lesions of PVM on red kidney bean; **E**, aggregate of PVM particles (**v**) and mitochondrion (**m**) in potato parenchyma cells. (D and E, Courtesy C. Hiruki)

Selected References

BAGNALL, R. H., R. H. LARSON, and J. C. WALKER. 1956. Potato viruses M, S and X in relation to interveinal mosaic of the Irish Cobbler variety. Wis. Agric. Exp. Stn. Res. Bull. 198. 45 pp.

BAGNALL, R. H., C. WETTER, and R. H. LARSON. 1959. Differential host and serological relationships of potato virus M, potato virus S and carnation latent virus. Phytopathology 49:435–442.

BAWDEN, F. C., B. KASSANIS, and H. L. NIXON. 1950. The mechanical transmission and some properties of potato paracrinkle virus. J. Gen. Microbiol. 4:210–219.

HIRUKI, C., E. POUNTNEY, and K. N. SAKSENA. 1974. Factors affecting bioassay of potato virus M in red kidney bean. Phytopathology 64:807–811.

KOWALSKA, A., and M. WAŚ. 1976. Detection of potato virus M and potato virus S on test plants. Potato Res. 19:131–139.

ROSENDAAL, A., and D. H. M. VAN SLOGTEREN. 1958. A potato virus identified with potato virus M and its relationship with potato virus S. Pages 20–36 in: F. Quak, J. Dijkstra, A. B. R. Beemster, and J. P. H. Van der Want, eds. Proc. Third Conf. Potato Virus Dis. 24–28 June, 1957. H. Veenman and Zonen, Lisse-Wageningen, The Netherlands. 282 pp.

TU, J. C., and C. HIRUKI. 1970. Ultrastructure of potato infected with potato virus M. Virology 42:238–242.

WETTER, C. 1972. Potato Virus M. No. 87 in: Descriptions of Plant Viruses. Commonw. Mycol. Inst., Assoc. Appl. Biol., Kew, Surrey, England.

(Prepared by C. Hiruki)

Potato Virus S

Potato virus S (PVS) occurs worldwide wherever the potato is grown. In the temperate zones, it is confined to the potato.

PVS was first detected not through symptoms but through serology during efforts to produce an antiserum to potato virus A.

Symptoms

PVS is virtually symptomless in most of the common potato

cultivars. Symptoms are slight deepening of veins and rugosity of leaves and possibly stunting and a more open type of growth. Some strains may cause mottling or bronzing in certain cultivars and, when severe, may cause necrotic spots on the upper surfaces. Older leaves in the shade may develop greenish spots instead of turning uniformly yellow. Controversy exists as to whether PVS alone consistently reduces yield, but losses of 10–20% have been reported.

Causal Agent

PVS particles are straight to slightly curved filaments, approximately 650×12 nm. Thermal inactivation is 55–60° C; dilution end point in crude potato sap is about 10^{-3}; and longevity in vitro at 20° C is about four days.

PVS is strongly antigenic, and for convenience and reliability in diagnosis, serology is generally used. Plants in the field are best tested just before flowering by sampling lower and middle leaves. Virus concentration may be low early or late in the season.

In tube tests, the precipitate is flocculent. The simple slide-agglutination test can be used under ideal conditions, but microprecipitin, bentonite, or latex flocculation or gel-diffusion tests are preferred.

Epidemiology

PVS is tuber-perpetuated, readily transmitted mechanically by infective sap, and reputed to be spread in nature primarily by contact with diseased plants. Certain strains are transmitted by the aphid *Myzus persicae* in a nonpersistent manner. Transmission tests with true seed have been negative.

Other Hosts

In *Nicotiana debneyi*, systemic vein clearing occurs after 20 days, spreading from the leaf tip toward the base (Fig. 80A). Later, leaves develop interveinal or veinbanding mottle at 20° C with a 16-hr day at 1,000 FC. Andean isolates give mild mosaic or are symptomless.

Solanum rostratum and *Saracha umbellata* (affected by some strains only) show necrotic spotting on inoculated leaves in 20 days, and later, on systemically invaded leaves. Andean strains do not cause necrotic spotting.

Datura metel, *Physalis philadelphica*, and *P. pubescens* are systemically infected without symptoms.

Capsicum annuum, *Lycopersicon esculentum*, *Nicandra physalodes*, *Nicotiana glutinosa*, *N. sylvestris*, *N. tabacum*, and *Physalis floridana* are immune to infection.

Local lesions occur in *Chenopodium album* and *C. amaranticolor* after 40 days (Fig. 80B and C), in *C. quinoa* (16–20 days), and in *Cyamopsis tetragonoloba* (6–10 days).

Most Andean strains (but not all) are systemic in *C. quinoa* and *C. amaranticolor*.

Resistance

Well-defined mature plant resistance is present in potato, so transmission must occur relatively early in the season if tubers are to become infected.

Certain cultivars, including Bintje, Katahdin, and Kennebec, are moderately resistant to reinfection. The cultivar Saco and some of its progeny have strong general resistance, which is inherited as a simple recessive; this resistance can be overcome by grafting, and plants carry the virus virtually without symptoms.

Resistance of the hypersensitive type is present in a cultivar of *Solanum tuberosum* ssp. *andigena*; this characteristic has been brought through several crosses with *S. tuberosum*. A similar type of resistance has been reported in the diploid *S. megistacrolobum*.

Control

1) PVS-free clones of susceptible cultivars have been established by meristem tip culture. In some regions these can be maintained relatively free from reinfection, but in other regions this is not so, possibly due to an insect vector. Indexing tubers and testing mother plants intended for production of stem cuttings, followed by release of elite seed, are the accepted practices.

2) Roguing is not effective.

Selected References

BAERECKE, M. L. 1967. Überempfindlichkeit gegen das S-Virus der Kartoffel in einen bolivianischen Andigena-Klon. Züchter 37:281–286.

BAGNALL, R. H., C. WETTER, and R. H. LARSON. 1959. Differential host and serological relationships of potato virus M, potato virus S, and carnation latent virus. Phytopathology 49:435–442.

BAGNALL, R. H., and D. A. YOUNG. 1972. Resistance to virus S in the potato. Am. Potato J. 49:196–201.

de BOKX, J. A., ed. 1972. Viruses of Potatoes and Seed-Potato Production. Pudoc, Wageningen, The Netherlands. 233 pp.

MacKINNON, J. P., and R. H. BAGNALL. 1972. Use of *Nicotiana debneyi* to detect viruses S, X, and Y in potato seed stocks, and relative susceptibility of six common varieties to potato virus S. Potato Res. 15:81–85.

SHEPARD, J. F., and L. E. CLAFLIN. 1975. Critical analyses of the principles of seed potato certification. Ann. Rev. Phytopathol. 13:271–293.

STACE-SMITH, R., and F. C. MELLOR. 1968. Eradication of potato viruses X and S by thermotherapy and axillary bud culture. Phytopathology 58:199–203.

Fig. 80. Potato virus S: **A,** systemic vein clearing in young leaves of *Nicotiana debneyi*; **B,** local lesions in young and old leaves of *Chenopodium amaranticolor*. (A, Courtesy R. H. Bagnall)

WETTER, C. 1971. Potato virus S. No. 60 in: Descriptions of Plant Viruses. Commonw. Mycol. Inst., Assoc. Appl. Biol., Kew, Surrey, England.

(Prepared by R. H. Bagnall)

Potato Virus T

Potato virus T (PVT) occurs in Peru, Bolivia, and probably elsewhere in the Andes.

Symptoms

PVT produces no obvious symptoms in several *tuberosum* and *andigena* types but may produce mild mottle or slight vein necrosis and chlorotic spots, and some *andigena* types may develop top necrosis after grafting. New shoots are symptomless but infected.

Causal Agent

PVT particles are flexuous filaments 640×12 nm, helically constructed with 3.4 nm pitch; their substructural detail differs from that of other viruses (Fig. 81). PVT is serologically related to, but distinct from, the apple stem grooving virus. Its thermal inactivation point is about $65°$C; dilution end point is 10^{-5}; and longevity in vitro is two to four days.

Epidemiology

PVT is readily transmitted mechanically by tubers and sap. It is not transmitted by the aphids *Myzus persicae* or *Macrosiphum euphorbiae*. It is seed-transmitted in *Nicandra physalodes*, *Datura stramonium*, and *Solanum demissum*, pollen-transmitted to seed but not to mother plants in *S. demissum*, and not pollen-transmitted in *N. physalodes* and *D. stramonium*.

Other Hosts

These include the tuber-bearing *S. demissum*, *S. stenotomum*, *S. chacoense*, *S. spegazzinii*, *S. vernei*, and *S. acaule*. PVT is readily transmitted by inoculation with sap and has a wide host range, infecting 43 of 56 species tested. Most solanaceous species remain symptomless or develop only mild symptoms.

Phaseolus vulgaris cv. Pinto and Prince are good local lesion hosts when heavily shaded after inoculation, later developing systemic necrosis followed by plant recovery.

Chenopodium quinoa, a good host for propagating the virus, is systemically infected.

C. amaranticolor is the best diagnostic host; it develops systemic distortion and necrosis.

Control

Little is know about diseases caused by PVT. Infected plants in stocks can be identified by visual inspection, by inoculation tests using suitable indicator plants, or by serological tests. Infected plants should be destroyed.

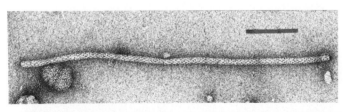

Fig. 81. Particle of potato virus T. Note unusual substructural detail. Bar represents 100 nm. (Courtesy L. F. Salazar and B. D. Harrison)

Selected References

SALAZAR, L. F., and B. D. HARRISON. 1977. Two previously undescribed potato viruses from South America. Nature 265:337–338.
SALAZAR, L. F., and B. D. HARRISON. 1978. Host range, purification and properties of potato virus T. Ann. Appl. Biol. 89:223–235.

(Prepared by L. F. Salazar and B. D. Harrison)

Andean Potato Mottle Virus

The Andean potato mottle virus (APMV) is present in Peru, Bolivia, probably throughout the Andean region at elevations of 2,000–4,000 m, and in Andean germplasm collections in Europe.

Symptoms

Primary symptoms are usually mild patchy mottle and, in sensitive cultivars, strong mottle, leaf deformation, systemic necrosis, and/or stunting. Strong secondary mottle is normal, but sensitive cultivars also show delayed emergence, leaf deformation, and severe stunting (Plate 64).

Causal Agent

APMV is a member of the cowpea mosaic virus (comovirus) group. Particles are isometric, approximately 28 nm in diameter, some full and some empty (Fig. 82C). Two types of proteins, small and large, have molecular weights of 20,800 and 40,100, respectively. The virus is strongly immunogenic and is serologically related to some members of the comovirus group. In *N. bigelovii* sap, the thermal inactivation point is $65–70°$C; longevity in vitro, four to five weeks; and dilution end point, $> 10^{-6}$.

Epidemiology

APMV produces symptoms and multiplies best under cool conditions. It is readily transmitted by contact between plants and probably also by animal and machinery movement. However, a beetle vector is possible because beetles characteristically vector the comovirus group.

Dissemination is through infected seed tubers.

Other Hosts

APMV can be transmitted mechanically only to solanaceous hosts.

Nicotiana bigelovii shows a mosaic of dark green blotches (Fig. 82A and B). Leaf tips develop necrotic areas and holes at $15–18°$C.

Lycopersicon chilense exhibits veinclearing, interveinal mosaic, chlorotic spotting, and sometimes epinasty.

Andean weeds, *Nicandra physalodes*, *Datura stramonium*, *Physalis peruviana*, and *L. pimpinellifolium* are susceptible to mechanical inoculation.

Control

Roguing is effective because APMV induces conspicuous symptoms.

Selected References

FRIBOURG, C. E., R. A. C. JONES, and R. KOENIG. 1977. Andean potato mottle, a new member of the cowpea mosaic virus group. Phytopathology 67:969–974.
FRIBOURG, C. E., R. A. C. JONES, and R. KOENIG. 1979. Andean potato mottle virus. No. 203 in: Descriptions of Plant Viruses. Commonw. Mycol. Inst., Assoc. Appl. Biol., Kew, Surrey, England. 4 pp.

(Prepared by C. E. Fribourg and R. A. C. Jones)

Andean Potato Latent Virus

Andean potato latent virus (APLV) is common throughout the Andean region at 2,000–4,000 m and occurs in collections of Andean germplasm in Europe.

Symptoms

Primary infection is often symptomless but may cause mosaics and/or chlorotic netting of minor leaf veins (Plate 65). Secondary infection normally causes mild mosaics, but chlorotic netting of minor leaf veins and rugosity may occur. Cool conditions favor symptom development.

Causal Agent

APLV resembles members of the turnip yellow mosaic group and is therefore considered a strain of eggplant mosaic virus. Particles are isometric and about 28 nm in diameter; some are full and some empty. The molecular weight of the protein subunits is 19,600–20,700. The virus is strongly immunogenic. Two serologically distinct strains are recognized (Col-Caj and Hu).

Dilution end point is 10^{-7}–10^{-8} and longevity in vitro is up to three weeks. Thermal inactivation ranges from 65 to 80° C, depending on the strain. Survival to 90° C has been claimed for Col-Caj in *Nicotiana glutinosa* sap.

Epidemiology

APLV produces symptoms and multiplies best under cool conditions. It is readily transmitted by contact between plants and probably also by animal and machinery movement. The potato flea beetle, *Epitrix*, is a vector of low efficiency. Transmission from infected potato plants to their tubers is erratic. The virus is transmitted at low frequency in potato true seed.

Other Hosts

APLV has been found occurring naturally only in potato. It can be transmitted mechanically to species of the Solanaceae, Chenopodiaceae, Cucurbitaceae, and Amaranthaceae, including the Andean crop plants *Chenopodium quinoa* and *Amaranthus edulis*.

In *N. bigelovii*, mosaic and characteristic netting of minor leaf veins develops with all isolates (Fig. 83), and most isolates also cause faint to distinct local lesions in inoculated leaves. More virulent isolates may also cause systemic necrotic flecking and starlike lesions.

N. clevelandii exhibits faint necrotic or chlorotic spots or rings on inoculated leaves with most isolates. Systemic symptoms are mosaic and necrotic or chlorotic netting of minor leaf veins.

Control

1) Clonal selection during initial multiplication of seed stocks is most effective. Tubers from infected plants often escape infection.

2) Roguing is effective only when conspicuous symptoms are present.

Fig. 82. Andean potato mottle virus in *Nicotiana bigelovii*, causing systemic necrotic areas and holes (**A**), or dark green blotchy mosaic (**B**). **C**, Virus particles in a partially purified preparation. The empty shells are dark colored. (A and B, Courtesy C. E. Fribourg; C and D, courtesy E. Lesemann)

Fig. 83. Andean potato latent virus mosaic and necrotic netting of minor veins in *Nicotiana bigelovii*. (Courtesy C. E. Fribourg)

3) Resistance has not been identified.

4) Applications of insecticides may help decrease spread when populations of *Epitrix* are high.

Selected References

FRIBOURG, C. E., R. A. C. JONES, and R. KOENIG. 1977. Host plant reactions, physical properties and serology of three isolates of Andean potato latent virus from Peru. Ann. Appl. Biol. 86:373–380.

GIBBS, A. J., E. HECHT-POINAR, R. D. WOODS, and R. K. McKEE. 1966. Some properties of three related viruses: Andean potato latent, Dulcamara mottle and Ononis yellow mosaic. J. Gen. Microbiol. 44:177–193.

JONES, R. A. C., and C. E. FRIBOURG. 1978. Symptoms induced by Andean potato latent virus in wild and cultivated potatoes. Potato Res. 21:121–217.

(Prepared by R. A. C. Jones and C. E. Fribourg)

Cucumber Mosaic Virus

Cucumber mosaic virus (CMV) causes chlorosis and a blistering mottle of leaves. Leaflet apices are usually elongate and margins distinctly wavy. Following mechanical inoculation, mild delimited chlorosis develops over a large part of the leaf surface as intense yellow flecks. Yellowing spreads slowly from the inoculated leaf to the petiole and is followed by complete collapse of the leaf. Symptoms successively develop on each higher leaf. The virus is recovered only from plant parts with symptoms. CMV on potato occurs naturally in England and Scotland. It is readily transmitted by rubbing with infective sap and by aphids in the nonpersistent manner.

CMV is usually not transmitted by tubers and has not become an economic problem.

Selected Reference

MacARTHUR, A. W. 1958. A note on the occurrence of cucumber mosaic virus in potato. Scott. Plant Breeding Stn. Rep. pp. 75–76.

(Prepared by W. J. Hooker)

Tobacco Mosaic Virus

Tobacco mosaic virus (TMV) is not a problem in potato production, having been reported once in field plants and in the wild in *S. commersonii*.

Mechanical inoculation of commercial cultivars results in localized infection, local lesions or blotches, drop of inoculated leaves, and rarely with systemic invasion in the current year. Usually infection is not tuberborne into the second year.

Green strains of TMV may be almost latent or cause symptoms suggestive of rugose mosaic virus, with leaflet twist, curl, and malformation. Yellow strains may exhibit striking yellowish flecking and leaflet malformation. Infected tubers develop multiple thin, weak stems, resembling those of witches' broom. Mosaic symptoms of TMV infection vary with the potato cultivar and virus isolate and may resemble those of potato viruses X, Y, or A.

Although resistance is present in commercially available cultivars of *S. tuberosum*, the virus is a potential threat. Susceptible commercial cultivars have apparently not been developed. Systemic infection readily transmitted through seed tubers occurs both within selections of *S. tuberosum* and also within certain wild species, including *S. acaule*.

Severity of reaction is greater at 24–28° C than at lower temperatures.

Potato plants may be killed when doubly infected with potato virus X and TMV.

Selected References

HANSEN, H. P. 1960. Tobacco mosaic virus carried in potato tubers. Am. Potato J. 37:95–101.

LUI, K. C., and J. S BOYLE. 1972. Intracellular morphology of two tobacco mosaic virus strains in, and cytological responses of, systemically susceptible potato plants. Phytopathology 62:1303–1311.

SIEMASZKO, J. 1961. Wirus mozaiki tytoniowej na ziemniakach. Biul. Inst. Hodowji Aklimatyzacii Roslin 5:13–18.

(Prepared by J. S. Boyle)

Potato Mop-Top Virus

Potato mop-top virus (PMTV) occurs in the Andean region of South America, in Northern and Central Europe, and probably also in other parts of the world where the powdery scab fungus, *Spongospora subterranea*, occurs.

Symptoms

PMTV causes yield decreases of up to 26% in sensitive cultivars and can have severe effects on tuber quality. Influenced by cultivar and environment, it causes variable symptoms. The most important of these are: 1) bright yellow (aucuba) markings: blotches, rings, or diagnostic V-shapes (chevrons), especially in lower leaves; 2) pale V-shaped markings in upper leaves; and 3) stunting of stems and shortening of internodes, involving some or all of the stems of a plant and giving the disease its name (Plate 66).

Primary infection from soil to tuber rarely spreads to the rest of the plant. Tubers are infected symptomlessly or with raised rings on the surface that, in sensitive cultivars, may be associated with necrotic arcs in the flesh (Plate 67). These arcs are induced at the virus invasion front in the tuber by a sudden drop in temperature, but they do not limit further virus spread. Secondary infection occurs only in some of the tubers produced by an infected plant. In sensitive cultivars, it may cause malformation, gross or fine superficial cracking, surface blotching, and/or necrotic arcs centered on the stolon (Plate 68).

Causal Agent

Particles are elongated, have a hollow core, and are normally defective because of terminal uncoiling of the protein helix. They are 18–20 nm wide, but of many lengths, frequently 250–300 or 100–250 nm. The helix pitch is 2.4–2.5 nm, and the molecular weight of the protein subunits is 18,500–20,000. The virus is moderately immunogenic and seems to be distantly related serologically to the tobacco mosaic virus.

Thermal inactivation is 75–80° C; dilution end point, 10^{-3}–10^{-5}; and longevity in vitro, up to 10 weeks.

Disease Cycle

PMTV survives inside *Spongospora subterranea* resting spores for several years and is transmitted to roots by zoospores.

The most important means of long-term survival in infested land is in dormant resting spores. Weed and other crop species in the Solonaceae and Chenopodiaceae, which are hosts of both virus and vector, are potential alternative natural hosts. The virus is introduced to uninfested fields when infected seed tubers carrying powdery scabs are planted. Also, short-distance movement may occur in infested soil transported on farm machinery or on the feet of farm animals. Healthy crops become infected when planted on infested land.

Epidemiology

Symptom production is especially sensitive to light and

temperature conditions. Systemic movement seems to occur through the xylem and not the phloem. PMTV causes distinct symptoms in potato only under cool conditions, and cold damp climates favor its spread by vector zoospores. In Northern Europe, the disease seldom occurs in areas with an annual rainfall of less than 760 mm.

Other Hosts

PMTV is known naturally only in potato but can be transmitted mechanically to species of the Solanaceae, Chenopodiaceae, and Aizoaceae.

In *Chenopodium amaranticolor*, diagnostic, spreading, concentric necrotic rings and local lesions form in shaded inoculated leaves at about 15° C. Infection is not systemic (Fig. 84A).

In *Nicotiana debneyi*, necrotic or chlorotic ringspots occur in inoculated leaves. Slow systemic invasion causes necrotic or chlorotic "thistle-leaf" line patterns (Fig. 84B). Symptom severity varies with different isolates. Systemic symptoms in *N. debneyi* bait seedlings are used to diagnose PMTV infestation in soil after air-drying.

Control

1) Roguing is effective only in stocks showing conspicuous symptoms.

2) No source of resistance is known, and heat treatment of tubers at 37° C does not eliminate PMTV from them.

3) Crop rotation does not control infestation of new land.

4) New infestation of fields can be decreased by treating seed tubers infected with both powdery scab and PMTV with formaldehyde or organomercurial fungicides.

5) Infection of healthy crops is decreased by treatment of infested fields with calomel, zinc compounds (e.g., ZnO), or by lowering pH to 5.0 by application of sulfur.

Selected References

CALVERT, E. L. 1968. The reaction of potato varieties to potato mop-top virus. Rec. Agric. Res. Minist. Agric., North Ireland 17:31–40.

COOPER, J. I., R. A. C. JONES, and B. D. HARRISON. 1976. Field and glasshouse experiments on the control of potato mop-top virus. Ann. Appl. Biol. 83:215–230.

JONES, R. A. C., and B. D. HARRISON. 1969. The behavior of potato mop-top virus in soil, and evidence for its transmission by *Spongospora subterranea* (Wallr.) Lagerh. Ann. Appl. Biol. 63:1–17.

JONES, R. A. C., and B. D. HARRISON. 1972. Ecological studies on potato mop-top virus in Scotland. Ann. Appl. Biol. 71:47–57.

KASSANIS, B., R. D. WOODS, and R. F. WHITE. 1972. Some properties of potato mop-top virus and its serological relationship to tobacco mosaic virus. J. Gen. Virol. 14:123–132.

(Prepared by R. A. C. Jones)

Tobacco Rattle Virus

The tobacco rattle virus (TRV) causes stem mottle of potato foliage and spraing of tubers on potatoes throughout Europe and in widely separated areas of North America and on hosts other than potato in Asia and South America. Stem mottle is rare in the field in North America.

Symptoms

Stem mottle primary symptoms occur on foliage following feeding by large numbers of viruliferous nematodes on emerging stems. Secondary systemic symptoms, more common in Europe, include some stunted stems with others that appear normal. Leaves may be mottled, puckered, distorted, or reduced in size (Plate 69, Fig. 85B). Some European strains cause a symptom resembling aucuba mosaic, with bright yellow, chevronlike stripes, arcs, and rings, which may be confused with other aucuba mosaics or with mop-top.

Spraing symptoms on tubers vary in type and severity with the degree of nematode infestation, time of infection, strain of the virus, potato cultivar, environmental conditions such as soil moisture and temperature, and type of infection (primary or secondary). Following nematode feeding directly on the tubers, external primary symptoms may be lacking or range from prominent concentric rings of alternating living and necrotic tissue to small necrotic flecks (Plate 70). Some cracking and various degrees of tuber malformation follow early season infection with some strains of TRV. Necrosis in the tuber flesh varies from prominent necrotic arcs and rings, usually but not always originating from surface lesions (Fig. 85A), to diffused necrotic flecks having no visible external involvement. Primary tuber symptoms are similar to those caused by mop-top virus except that the latter tend to be concentrated on the tuber surface.

Secondary tuber symptoms tend to be malformation and internal flecking. Many tubers on stem mottle plants contain TRV but remain symptomless. Tubers with mild necrosis tend to produce more stem mottle plants than do tubers with severe necrosis. Certain strains of TRV tend to produce mostly stem mottle plants, whereas other strains cause mostly spraing.

Stem mottle is more important in seed production, and spraing is more serious with table stock.

Causal Agent

TRV is a rod-shaped virus, 17–25 nm in diameter, with long infectious rods (about 180–210 nm), which do not code for coat protein synthesis, and short noninfectious rods (about 45–115

Fig. 84. Potato mop-top virus: **A,** in *Chenopodium amaranticolor*, showing diagnostic concentric, necrotic ring, local lesions; **B,** systemic necrotic "thistle leaf" symptoms in *Nicotiana debneyi* bait seedlings. (Courtesy R. A. C. Jones)

nm), which do code for coat protein synthesis (Fig. 85C). Protein-coated long and short rods are produced only when the inoculum contains particles of both lengths. Many strains of TRV are known and may be differentiated serologically.

Isolation of TRV from potato is usually difficult, and difficulty increases with time after harvest. TRV is isolated from soil by mechanically inoculating leaves of test plants with roots of bait plants (tobacco or cucumber) triturated in buffer.

Epidemiology

Incidence of spraing is highest in sandy soils. Twelve species of stubby root nematode (*Trichodorus* or *Paratrichodorus*) transmit TRV; however, not all of these feed on potato. *Trichodorus* spp. generally occur in sandy, open-textured soils and have been found at depths of 80–100 cm. Populations of *Trichodorus* have remained infective in moist soil maintained in the laboratory for as long as five years in the absence of a host. The virus is not retained through the egg or the molt.

Primary tuber infection follows nematodes feeding directly on tubers. Systemic infection of tubers from root feeding sites has not been demonstrated. Systemic movement of TRV occurs most often at 20–25° C and is sometimes enhanced by pruning.

Although the disease can be transmitted through seed tubers, transmission in this manner is infrequent, and the virus is ultimately self-eliminating. Nonviruliferous populations of the *Trichodorus* rarely acquire the virus from infected potato plants, and tuber dissemination of TRV from infected to virus-free fields has not been demonstrated. Opinions differ as to the importance of tuber dissemination. The virus can be seed-transmitted in other hosts. Movement of soil infested with viruliferous nematodes is probably the most effective means of dissemination.

Other Hosts

TRV can infect more than 400 monocotyledonous and dicotyledonous species in over 50 families. Often it does not become systemic, remaining in roots of plants that show no foliage symptoms.

In *Chenopodium amaranticolor,* the virus produces necrotic local lesions, some spreading, but the virus is not systemic. The host is used for diagnosis and for local lesion assay.

In *Cucumis sativus,* chlorotic or necrotic local lesions occur. This is a diagnostic host, used for local lesion assay, and is a good bait plant as roots become infected from nematode feeding.

On *Nicotiana clevelandii,* inoculated leaves are symptomless or have chlorotic or necrotic lesions. Systemically infected leaves develop few symptoms or variable amounts of necrotic flecking and distortion. This is a diagnostic host, also used for increasing virus titer.

Necrotic local lesions are formed on *Phaseolus vulgaris* in one to three days. The disease is not systemic.

N. tabacum cultivars White Burley, Samsum NN, and others often have local lesions and serve for local lesion assay. They are good bait plants as roots become infected.

Control

1) Cultivars resistant to spraing in one field may not be resistant in other fields, even in the same country.

2) Soil fumigants such as 1,3-dichloropropene, related C_3 hydrocarbons, and dazomet and nonvolatile organophosphate and carbamate chemicals such as phenamiphos, oxamyl, aldicarb, carbofuran provide effective control. Effectiveness of chemicals varies with location.

3) Soil fumigation and consecutive plantings of potatoes may almost eliminate the vector from the soil because potato is a poor host for *Trichodorus* spp.

4) Rotational or cover crops have been considered of little use because of the wide host range of the vector and virus. Incidence of spraing may be reduced in potatoes following cover crops such as *Hordeum* spp., which are resistant to TRV but are good hosts for the vector.

Selected References

HARRISON, B. D. 1968. Reactions of some old and new British potato cultivars to tobacco rattle virus. Eur. Potato J. 11:165–176.

HARRISON, B. D. 1970. Tobacco rattle virus. No. 12 in: Descriptions of Plant Viruses. Commonw. Mycol. Inst., Assoc. Appl. Biol., Kew, Surrey, England.

HARRISON, B. D. 1977. Ecology and control of viruses with soil-inhabiting vectors. Ann. Rev. Phytopathol. 15:331–360.

HARRISON, B. D., and R. D. WOODS. 1966. Serotypes and particle dimensions of tobacco rattle viruses from Europe and America. Virology 28:610–620.

TAYLOR, C. E. 1971. Nematodes as vectors of plant viruses. Pages 185–211 in: B. M. Zuckerman, W. F. Mai, and R. A. Rohde, eds. Plant Parasitic Nematodes. II. Cytogenetics, Host-Parasite Interactions, and Physiology. Academic Press, New York. 347 pp.

VAN HOOF, H. A. 1972. Soilborne viruses. Pages 57–64 in: J. A. de Bokx, ed. Viruses of Potatoes and Seed-Potato Production. Pudoc, Wageningen, The Netherlands. 233 pp.

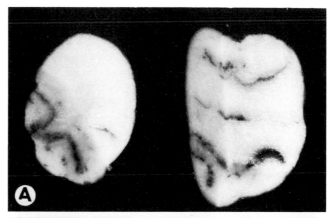

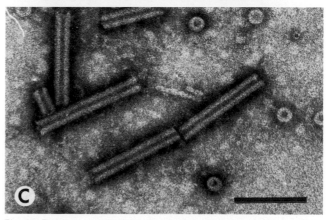

Fig. 85. Tobacco rattle virus: **A,** necrotic arcs and rings in tubers; **B,** severe stunting and leaf distortion; **C,** virus particles, long and short rods, and fragments of rods in cross section (bar represents 100 nm). (B, Courtesy D. P. Weingartner; C, courtesy R. Stace Smith)

WALKINSHAW, C. H., and R. H. LARSON. 1959. Corky ringspot of potato. A soil-borne virus disease. Wis. Agric. Exp. Stn. Res. Bull. 217. 31 pp.

(Prepared by D. P. Weingartner)

Potato Yellow Dwarf Virus

Potato yellow dwarf virus (PYDV) occurs in Canada and in the United States in Michigan, New York, and Wisconsin. Disease outbreaks have not been reported for almost 40 years.

Symptoms

Vines from infected seed pieces are dwarfed and brittle, and the entire plant has a yellowish cast. Leaflet margins roll upward, but the longitudinal axis of the leaflet curves downward (Plate 71). Pith necrosis of stems is common, appearing shortly after foliage chlorosis. Necrosis beginning near the growing point may eventually extend the length of the stem (Fig. 86).

Tubers are usually few, small, and deformed. A dark brown necrosis, which may easily be confused with heat necrosis, occurs throughout the tuber, with only the xylem elements apparently unaffected (Fig. 86). Failure of infected tubers to germinate, resulting in widespread stand reduction, is characteristic.

Causal Agent

PYDV is bacilliform, with particles measuring about 380 × 75 nm. Virions are closely associated with the nucleus of infected cells. Approximately 20% of the virus is lipid. Of the four major structural proteins (with molecular weights of 22, 33, 56, and 78 × 10^3), the largest is a glycoprotein. The virus contains single-stranded ribonucleic acid (molecular weight 4.6 × 10^6).

Infectivity of PYDV in *Nicotiana rustica* sap is retained for 2.5–12 hr at 23–27°C. Dilution end point is usually about 10^{-3} and thermal inactivation, about 50°C. PYDV does not withstand desiccation in *N. rustica* leaves nor prolonged storage in frozen leaves. Purification of PYDV is difficult because the virus is quite labile.

Epidemiology

PYDV, the only known virus borne by leafhoppers that is also mechanically transmissible, consists of two closely related forms, one transmitted by *Aceratagallia sanguinolenta* but not by *Agallia constricta* and another transmitted with the inverse vector relationship. *Agallia quadripunctata* has been reported to transmit both forms. PYDV is propagative in the vector. No transmission through true seed has been recorded. PYDV is carried from one generation to the next in potato tubers.

High temperatures enhance vine symptoms and reduce plant emergence from infected tubers, whereas low temperatures increase plant or sprout emergence and suppress vine symptoms.

Other Hosts

In addition to solanaceous plants, vectors have transmitted the virus to members of Compositae, Cruciferae, Labiatae, Leguminosae, Polygonaceae, and Scrophulariaceae.

N. rustica and *N. glutinosa* can be mechanically inoculated by rubbing.

In *N. rustica*, primary lesions occur, followed by systemic invasion. This host serves for virus assay and propagation.

Chrysanthemum leucanthemum var. *pinnatifidum*, oxeye daisy, is the principal source of infection for potato crops.

Control

1) Plant certified seed produced in areas where PYDV is not found.

2) Plant certified seed produced as far as possible from clover fields in infested areas to avoid the clover leafhopper, *Aceratagallia sanguinolenta*, which can harbor the virus through the winter. Table stock potatoes should not be planted adjacent to clover fields in infested areas.

3) Plant tolerant cultivars. Cultivars shown to be field-tolerant are Chippewa, Katahdin, Russet Burbank, and Sebago.

Selected References

BLACK, L. M. 1937. A study of potato yellow dwarf in New York. N. Y. Agric. Exp. Stn., Cornell, Mem 209. 23 pp.

BLACK, L. M. 1970. Potato yellow dwarf virus. No. 35 in: Descriptions of Plant Viruses. Commonw. Mycol. Inst., Assoc. Appl. Biol., Kew, Surrey, England.

HSU, H. T., and L. M. BLACK. 1973. Polyethylene glycol for purification of potato yellow dwarf virus. Phytopathology 63:692–696.

MacLEOD, R., L. M. BLACK, AND F. H. MOYER. 1966. The fine structure and intracellular localization of potato yellow dwarf virus. Virology 29:540–552.

MUNCIE, J. H. 1935. Yellow dwarf disease of potatoes. Mich. Agric. Exp. Stn. Spec. Bull. 260. 18 pp.

WALKER, J. C., and R. H. LARSON. 1939. Yellow dwarf of potato in Wisconsin. J. Agric. Res. 59:259–280.

(Prepared by H. Darling and S. Slack)

Alfalfa Mosaic Virus

Alfalfa mosaic virus (AMV) is found worldwide but is generally considered of little economic importance in potatoes.

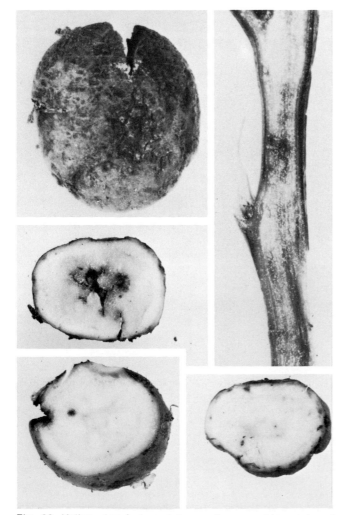

Fig. 86. Yellow dwarf virus: tuber surface cracking, internal necrosis, and pith necrosis of stem. (Courtesy J. H. Muncie)

Symptoms

AMV may induce predominantly calico symptoms on foliage (Plate 72) or necrotic symptoms in tubers. Calico symptoms are pale to bright mottling or blotching of potato leaflets, with all or large sectors of leaflets yellow. The plant may be slightly stunted. Leaflet necrosis is common and may extend into the stems and to the tubers.

Tuber necrosis, usually visible by harvest, begins just beneath the epidermis at the stolon attachment and later spreads throughout the tuber, leaving scattered dry, corky areas. Tubers may be misshapen, cracked, and few per plant. Disease severity varies with virus strains and potato cultivars.

Causal Agent

AMV consists of several differently sedimenting nucleoprotein species (Fig. 87), each with a width of about 18 nm and containing about 18% ribonucleic acid (RNA). The larger particles appear to be bacilliform, with approximate lengths of 58, 49, 38, and 29 nm, whereas the smallest particle, 19 nm in diameter, is probably icosahedral. Molecular weights for the encapsidated RNAs are about 1.3, 1.1, 0.9, and 0.3×10^6. A single structural protein with a molecular weight value between 24,500 and 32,600 has been reported. The largest three nucleoprotein components or all four RNA molecules are necessary for infection. The structural or coat protein can be substituted for the smallest RNA molecule to initiate infection. AMV is moderately immunogenic, and no serological relationship to other viruses has been found.

Thermal inactivation is usually between 60 and 65°C but may range from 50 to 75°C. The dilution end point is generally in the vicinity of 10^{-3}, and longevity in vitro is from 4 hr to four days. Infectivity in leaves persists at −18°C for more than a year.

Epidemiology

Mechanical inoculation with sap by rubbing is generally effective. AMV is transmitted in the styletborne or nonpersistent manner by as many as 16 aphid species, including *Myzus persicae*. In potato, AMV may be transmitted by tuber-core grafting and can be carried from season to season in tubers. Although AMV is seed-transmitted in some alfalfa varieties and in chili pepper, transmission through true potato seed has not been demonstrated.

Other Hosts

Natural infection is known for 47 plant species in 12 families; over 300 species have been experimentally infected. AMV is highly variable, with numerous strains causing different reactions on test plants.

Phaseolus vulgaris (French bean) and *Vigna sinensis* generally show local lesions and/or systemic infection and serve as diagnostic and assay plants.

Vicia faba and *Pisum sativum* exhibit black, necrotic, local lesions and stem necrosis; plants die.

In *Chenopodium amaranticolor* and *C. quinoa,* local lesions and systemic chlorotic and necrotic flecks are produced.

Nicotiana tabacum is a good propagation species, with necrotic or chlorotic local lesions and systemic mottle.

Control

1) Potatoes should not be planted adjacent to known reservoir hosts such as alfalfa or clover, especially when cultural practices promote aphid movement into the potato crop.

2) Volunteer alfalfa, clover, or potato plants that may serve as an initial source of inoculum should be removed from a potato field.

3) Visibly infected potatoes should be rogued during the growing season.

4) Certified seed should be planted each year.

5) Spraying to reduce vector populations is of dubious value because of the short time required for transmission.

Selected References

BEEMSTER, A. B. R., and A. ROZENDAAL. 1972. Potato viruses: Properties and symptoms. Pages 115–143 in: J. A. de Bokx, ed. Viruses of Potatoes and Seed-Potato Production. Pudoc, Wageningen, The Netherlands. 233 pp.

BLACK, L. M., and W. C. PRICE. 1940. The relationship between viruses of potato calico and alfalfa mosaic. Phytopathology 30:444–447.

BOL, J. F., L. van VLOTEN-DOTING, and E. M. J. JASPARS. 1971. A functional equivalence of top component *a* RNA and coat protein in the initiation of infection by alfalfa mosaic virus. Virology 46:73–85.

BOS, L., and E. M. J. JASPARS. 1971. Alfalfa mosaic virus. No. 46 in: Descriptions of Plant Viruses. Commonw. Mycol. Inst., Assoc.

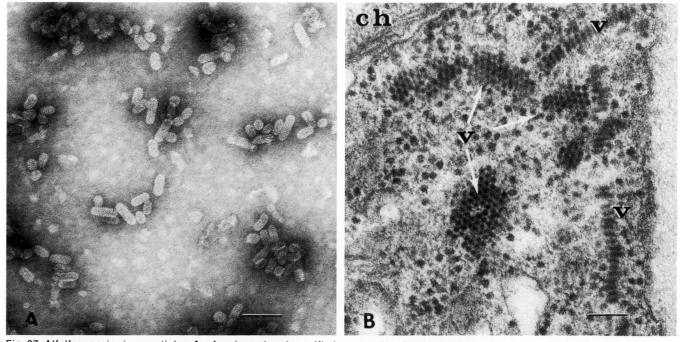

Fig. 87. Alfalfa mosaic virus particles: **A,** of various sizes in purified preparation, fixed in 2% glutaraldehyde and negatively stained in 1% phosphotungstate; **B,** one type of particle aggregation occurring in infected tissue (ch = chloroplast, v = virus). Note the pronounced swelling of virus particles in A versus those in B. Bars represent 100 nm. (Courtesy G. A. de Zoeten and G. Gaard)

Appl. Biol., Kew, Surrey, England.

CERVANTES, J., and R. H. LARSON. 1961. Alfalfa mosaic virus in relation to tuber necrosis in the potato variety Red La Soda. Wis. Agric. Exp. Stn. Res. Bull. 229. 40 pp.

HULL, R. 1969. Alfalfa mosaic virus. Adv. Virus Res. 15:365–433.

OSWALD, J. W. 1950. A strain of the alfalfa-mosaic virus causing vine and tuber necrosis in potato. Phytopathology 40:973–991.

(Prepared by S. Slack)

Potato Aucuba Mosaic Virus

Potato aucuba mosaic virus (PAMV) is found worldwide but is not common. It is also called tuber blotch.

Symptoms

These depend both on virus strain and potato cultivar (Fig. 88 A and B). They include bright yellow spots, mostly on the lower leaves; deformation and stunting without yellow spots; mosaic and top necrosis; and necrosis and sometimes deformation of the tubers (tuber blotch) (Fig. 88C).

Foliage symptoms may be lacking in the second year after infection and later. Tuber necrosis generally develops during storage and its development is more severe at higher temperatures (20–24° C) than at lower ones. Necrosis of the tubers can be both on the surface (brown patches and sunken brown areas) and in the flesh.

Fig. 88. Potato aucuba mosaic virus: **A**, bright yellow mosaic mottle; **B**, deformation; **C**, tuber necrosis; **D**, necrosis of inoculated leaves and systemic leaf necrosis in pepper. (Courtesy J. A. de Bokx)

Originally, PAMV (potato virus G) and potato tuber blotch virus (potato virus F) were described as different viruses. Now the latter is considered to be a strain of PAMV. Many strains are known and can best be differentiated by using various potato cultivars.

Causal Agent

PAMV has filamentous particles, 580 nm long and 11–12 nm wide. The virus is transmitted mechanically and by aphids in the nonpersistent manner (styletborne) when aided by "helper" viruses, including potato viruses Y or A.

The virus is strongly immunogenic. Antisera can be used for identification by the precipitin test.

Other Hosts

On *Capsicum annuum*, brown, irregular concentric local lesions appear after 8–10 days, followed by systemic symptoms such as vein clearing, deformation, and severe necrosis (Fig. 88D) and sometimes by complete killing of the plants. PAMV can be differentiated from mop-top virus, which produces similar symptoms on potato, because the latter does not infect *C. annuum* systemically.

Nicotiana glutinosa exhibits mottle and vein banding.

In *N. tabacum*, PAMV usually produces symptomless systemic infection.

Some strains produce small round yellow spots on lower leaves of *Lycopersicon esculentum*.

Control

Remove infected plants from seed fields.

Selected References

BEEMSTER, A. B. R., and A. ROSENDAAL. 1972. Potato viruses: Properties and symptoms. Pages 115–143 in: J. A. de Bokx, ed. Viruses of Potatoes and Seed-Potato Production. Pudoc, Wageningen, The Netherlands. 233 pp.

KASSANIS, B., and D. A. GOVIER. 1972. Potato aucuba mosaic virus. No. 98 in: Descriptions of Plant Viruses. Commonw. Mycol. Inst., Assoc. Appl. Biol., Kew, Surrey, England.

MUNRO, J. 1960. The reactions of some potato varieties and seedlings to potato virus F. Am. Potato J. 37:249–256.

SMITH, K. M. 1972. A Textbook of Plant Virus Diseases, 3rd ed. Longman, London. 684 pp.

(Prepared by A. B. R. Beemster)

Tobacco Ringspot Virus

So far, Andean potato calico caused by tobacco ringspot virus (TRSV) has been confirmed only in Peru, but preliminary tests indicate its presence in other Andean countries.

Symptoms

Bright yellow areas on the margins of middle and upper leaves gradually increase in size to form large patches or even to affect the whole leaf (Plate 73). Most of the plant foliage may eventually turn yellow without stunting or leaf deformation (Plates 73 and 74). In experimentally inoculated potato, the primary reactions are local and systemic necrotic spots, ringspots, and sometimes systemic necrosis.

Causal Agent

Particles are isometric, about 28 nm in diameter. Purified preparations show empty particles and particles with infectious or noninfectious nucleoprotein. The virus is highly immunogenic. TRSV occurs in nature as six well-characterized, serologically related strains. The Andean potato calico strain is serologically related, but not identical, to the others. In tobacco sap its thermal inactivation point is 55–60° C; longevity in vitro, 9–11 days; and dilution end point, $10^{-4}–10^{-5}$. TRSV is similar in certain respects to potato black ringspot virus.

Disease Cycle

The virus is probably introduced to uninfested soil when infected seed tubers are planted. Seedlings from infected seed of some weed species may also act as virus reservoirs. Transmission to potato plants has not been investigated, but the virus is transmitted in other crops by the nematode, *Xiphinema americanum*; thrips, *Thrips tabaci*; the tobacco flea beetle, *Epitrix hirtipennis*; and other vectors. However, sporadic field occurrence of calico fits the pattern of spread by a nematode vector.

Epidemiology

Cool temperatures favor disease development in potato plants. High soil moisture may favor the nematode.

Other Hosts

TRSV transmits mechanically to at least 38 genera in 17 families, usually inducing symptoms of chlorotic or necrotic spotting.

In *Nicotiana tabacum*, it produces local and systemic chlorotic and necrotic ringspots and line patterns, followed by systemic symptomless infection (Fig. 89A).

Vigna sinensis exhibits reddish necrotic lesions in inoculated leaves, followed by systemic apical necrosis (Fig. 89B).

In *Cucumis sativus*, symptoms are chlorotic or necrotic lesions in inoculated leaves, systemic chlorotic spots, mottling, and apical distortion.

Control

1) Roguing infected plants is effective because the disease is sporadic and susceptible cultivars show conspicuous symptoms.

2) TRSV is sometimes controlled in other crops by treating infested fields with nematicides.

Selected References

FRIBOURG, C. E. 1977. Andean potato calico strain of tobacco ringspot virus. Phytopathology 67:174–178.
GOODING, G. V., Jr. 1970. Natural serological strains of tobacco ringspot virus. Phytopathology 60:708–713.
SALAZAR, L. F., and B. D. HARRISON. 1978. The relationship of potato black ringspot virus to tobacco ringspot and allied viruses. Ann. Appl. Biol. 90:387–394.
STACE-SMITH, R. 1970. Tobacco ringspot virus. No. 17 in: Descriptions of Plant Viruses. Commonw. Mycol. Inst., Assoc. Appl. Biol., Kew, Surrey, England.

(Prepared by C. E. Fribourg)

Tomato Black Ring Virus

Tomato black ring virus (TBRV) affects potato in Northern and Central Europe but normally is uncommon.

Symptoms

Primary systemic symptoms, necrotic spots and rings on leaves, develop in only a few plants. Often plants are symptomless the first year of infection. In the second year, chronic symptoms develop; tip leaves may be cupped (spoon-shaped) and slightly misshapen, with necrotic spotting and plants somewhat stunted (bouquet) (Fig. 90). In the third year, plants develop only chronic symptoms. Bright yellow leaf markings are produced by the pseudo-aucuba strain. Tubers are infected symptomlessly, and some tubers from diseased plants are healthy. Although loss of yield in individual plants with secondary symptoms averages 20–30%, this is not a major disease because it occurs only sporadically.

Causal Agent

Particles are isometric and about 30 nm in diameter. Some are full and some empty. The virus is strongly immunogenic. In tobacco sap, thermal inactivation is 60–65°C; dilution end point, 10^{-3}–10^{-4}; and longevity in vitro, two to three weeks.

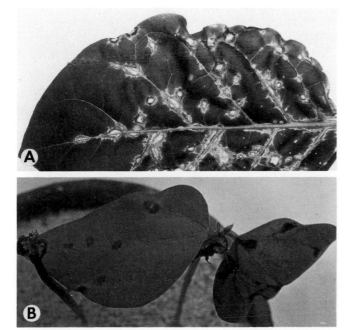

Fig. 89. Tobacco ringspot virus: **A**, systemic symptoms in tobacco leaf; **B**, local necrotic lesions and systemic apical necrosis in cowpea (*Vigna sinensis*). (Courtesy C. E. Fribourg)

Fig. 90. Tomato black ring virus systemic necrotic spotting in naturally infected Majestic variety. (Courtesy B. D. Harrison)

Disease Cycle

TBRV is soilborne and transmitted by *Longidorus* spp. nematodes. *L. attenuatus* transmits the bouquet strain and *L. elongatus* the pseudo-aucuba strain. The virus is not retained through the moult, and infective nematodes kept in fallow soil lose infectivity after about nine weeks.

The most important means of survival in infested land is in infected weeds and their seeds. TBRV is seed-transmitted in at least 24 species in 15 families. Seeds also serve as vehicles for dissemination of the virus to new sites, as do infected seed tubers. Short distance movement may occur by nematodes in infested field soil. Healthy crops become infected when planted on infested land.

Epidemiology

Cool climates favor development of disease in potato plants. Activity of the nematode vector is favored by high soil moisture, with resultant spread to healthy potato plants.

Other Hosts

TBRV can be mechanically transmitted to a wide range of hosts in at least 29 families and naturally infects many weed and crop plants.

Chenopodium amaranticolor and *C. quinoa* exhibit chlorotic or necrotic spots and rings in inoculated leaves and systemic necrosis or chlorotic mottle.

Nicotiana rustica and *N. tabacum* show necrotic or chlorotic spots or rings in inoculated leaves and systemic spots, rings, and line patterns with variable amounts of necrosis. Leaves produced later appear normal but contain virus.

Cucumis sativus is a useful bait plant for virus detection in soil samples, producing chlorotic or necrotic local lesions and systemic mottling. Leaves produced later may develop enations.

Control

1) Roguing is effective if seed stocks are not later reinfected by being planted in infested land.

2) Nematicide soil treatments decrease incidence of infection when healthy crops are planted.

Selected References

BERCKS, R. 1962. Serologische Überkreuzreaktionen zwischen Isolaten des Tomatenschwarzringflecken-Virus. Phytopathol. Z. 46:97–100.

HARRISON, B. D. 1958. Relationship between beet ringspot, potato bouquet and tomato black ring viruses. J. Gen. Microbiol. 18:450–460.

MURANT, A. F. 1970. Tomato black ring virus. No. 38 in: Descriptions of Plant Viruses. Commonw. Mycol. Inst., Assoc. Appl. Biol., Kew, Surrey, England.

(Prepared by R. A. C. Jones and C. E. Fribourg)

Potato Yellow Vein Virus

Vein yellowing, caused by the potato yellow vein virus (PYVV), is very common in the highlands of Ecuador and southern Colombia. Apparently PYVV is poorly invasive in the plant; tuber transmission is not regular and tuberborne infection may be very slow in symptom expression.

Newly developed symptoms are bright yellow veins and, in some cultivars, interveinal yellowing. Later the leaves become yellow and the veins may turn green (Plate 75). These colors are most distinct in leaves that had expanded before the onset of the disease. Affected plants are spectacular because of the intensely bright yellow color, which remains distinct throughout the life of affected leaves. Leaves with symptoms are somewhat rougher than are leaves that appear healthy. Some rugosity and necrotic spotting may also develop. Tubers are deformed, with eyes protruding as knobs suggestive of second growth. Yields may be reduced by 50%.

The PYVV particle is isometric and 26 nm in diameter (Salazar, L. F., and B. D. Harrison. 1980. Personal communication). The vector in nature has not been identified. The disease is graft-transmitted and can be mechanically transmitted to *Datura stramonium* only with difficulty.

Selected References

DIAZ, MORENO, J. 1965. El virus del amarillamiento de las papas. Cienc. Nat. (Quito) 8:25–37.

VEGA, J. G. 1970. Transmissión, purificación y caracterización del agente causal del "amarillamiento de venas" en papa. Tesis, M. S. Universidad Nacional Instituto Colombiano Agropecuario. Bogatá. 47 pp.

SMITH, K. M. 1972. A Textbook of Plant Virus Diseases, 3rd ed. Academic Press, New York. pp. 427–428.

(Prepared by W. J. Hooker)

Tobacco Necrosis Virus

Tobacco necrosis virus (TNV) occasionally affects the cultivars Duke of York (Eersteling) in the Netherlands and Sieglinde in Italy.

Symptoms

Only tubers react to infection (Fig. 91). The skin of recently lifted tubers shows dark brown lesions with radial or reticular cracks. Parallel or star-shaped cracks may resemble scab lesions. Superficial lesions are circular or bandlike, and light brown patches about the same size as the radial cracks occur. Blisters, sometimes visible at harvest, may develop during storage and later become sunken, covering most of the tuber surface.

Causal Agent

TNV has a wide host range, and little is known about the disease of potato. Inoculation of Duke of York tubers with TNV has been unsuccessful. Diseased tubers produce a healthy or nearly healthy progeny.

TNV rarely becomes systemic, is transmitted naturally to plant roots by zoospores of *Olpidium brassicae*, infects both monocotyledonous and dicotyledonous plants, and after mechanical inoculation typically produces local lesions on leaves.

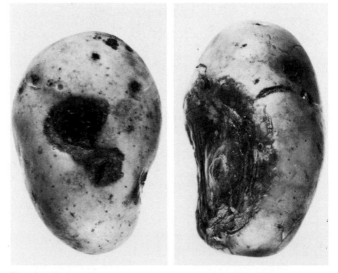

Fig. 91. Tobacco necrosis virus: early and late symptoms on tubers. (Courtesy D. Peters)

Selected References

KASSANIS, B. 1970. Tobacco necrosis virus. No. 14 in: Descriptions of Plant Viruses. Commonw. Mycol. Inst., Assoc. Appl. Biol., Kew, Surrey, England.

NOORDAM, D. 1957. Tabaksnecrosevirus in samenhang met een oppervlakkige aantasting van aardappelknollen. Tijdschr. Plantezieken 63:237–241.

(Prepared by D. Peters)

Deforming Mosaic

Relatively little is known about this disease, although it is economically important in Argentina.

Secondary symptoms appearing soon after plant emergence are severe mosaic and irregular occurrence of yellowish green patterns. Leaves are distorted, and leaf surfaces are rough, with interveinal tissue extending above the veins. Later in the season, severity of symptoms lessens and infection may be masked.

Tubers are symptomless, internally and externally. Tuber transmission of the virus is not consistent, and up to a third of the tubers from infected plants may produce healthy plants.

The disease is graft-transmitted to *Solanum demissum* and *S. chacoense* and then readily transmitted to potato. It has not yet been transmitted mechanically nor by aphids. The potato cultivar Huinkul is rapidly infected in the field, with losses of 20% or more (Fig. 92). Others, Kennebec and Pontiac, usually do not show symptoms under similar exposure.

Early roguing of diseased plants in seed fields is recommended.

Selected Reference

CALDERONI, A. 1965. An unidentified virus of deforming mosaic type in potato varieties in Argentina. Am. Potato J. 42:257 (Abstr.).

(Prepared by W. J. Hooker)

Tomato Spotted Wilt Virus

The tomato spotted wilt virus (TSWV) occurs worldwide but is more frequent in subtropical and temperate regions of the African, American, and Australian continents than in Europe and Asia. Losses in potato can be extensive.

Symptoms

Primary symptoms are necrotic spotting of leaves, stem necrosis, death of the top of one or more stems, and occasionally death of the whole plant (Plate 76). More frequently, however, only the upper parts die.

Local chlorotic or necrotic lesions may appear at points of thrip feeding (Fig. 93A and D). Pale green upper growth may precede the onset of systemic necrotic spotting. Leaf lesions appear as single necrotic rings around a central green area with or without a central dot, as concentric necrotic rings interspaced with green tissues, or as solid necrotic spots with concentric zonation somewhat resembling *Alternaria* leaf spot. Necrotic lesions appear also in the petioles and veins and in the stem (cortex and/or pith). Axillary buds may sprout in necrotic plants with chlorotic concentric ringspots.

Tubers produced by infected plants may appear normal or they may be malformed, with cracks and internal rusty or dark necrotic spots. Spots are visible when the tuber is cut or may be visible through the skin, sometimes as concentric patterns (Fig. 93B and C).

Secondary symptoms on shoots from infected tubers may include necrosis, early death, varying degrees of stunting, or a rosette type of growth with coarse, dark green leaves. Leaves may show necrotic spotting or chlorotic concentric ringspots.

Many diseased plants survive with minimal yield of tubers, which are generally small and malformed.

Causal Agent

TSWV consists of isometric particles 70–100 nm in diameter (Fig. 93F). They contain ribonucleic acid and 19% lipids. Particles occur singly or in clusters inside membrane-bound structures that correspond to cisternae in the endoplasmic reticulum or nuclear envelope. The virus membrane may be partly of cellular origin. Virus particles occur in leaf, stem, and root tissues and are present in all types of leaf cells except tracheids.

TSWV is unstable in plant extracts but more stable with buffers near pH 7 containing a reducing system such as sodium sulfite, thioglycollate, or cysteine. Inactivation is at 50° C for 10 min; longevity in vitro is 2–5 hr; infectivity is lost at relatively low dilutions. TSWV is serologically active.

The virus occurs in nature as a series of strain complexes with certain strains predominating in potato and thus differing, in part, from those affecting other crops. Some strains tend to become localized in potato leaf tissues when inoculated individually but may invade the plant systemically in the presence of others that act in an auxiliary capacity.

Epidemiology

The spotted wilt virus is vectored by thrips: *Thrips tabaci, Frankliniella schultzei, F. fusca, and F. occidentalis.* The first two are involved in spread in potato. The virus is acquired only by immature stages of thrips and transmitted only by adults. Therefore, symptoms of a new infection transmitted from other potato plants by thrips may not appear for several weeks while the vector is developing.

Transmission is unlikely through true seed of potato, although some transmission through seed of other plants has been reported.

Virus perpetuation in either normal or malformed tubers may reach 30–40% but generally does not exceed 5%. Some tubers from a diseased hill may be virus-free, and some buds from a diseased tuber carry the virus whereas others do not.

The spotted wilt virus can be transmitted mechanically with relative ease if infected leaves are triturated with a reducing substance.

Mechanical transmission in potato apparently does not occur

Fig. 92. Deforming mosaic in cultivar Huinkul. (Courtesy A. V. Calderoni)

naturally. Potato, more resistant to mechanical inoculation than most other suscepts, is most effectively inoculated on the young tips of plants near the flowering stage with virus from potatoes rather than from other hosts.

Spotted wilt in potato is seasonal. Spread generally occurs in late spring and early summer, occasionally later, and sometimes erratically.

Most field infection occurs as viruliferous vectors move into potato from outside sources. Field incidence may be patchy, especially during early growth or when incidence is low. High densities of the vector, even if present for only short periods, seem more necessary for widespread potato infection than for infection of other crops such as tobacco and tomatoes.

In Australia, thrip populations increase when the rainfall is satisfactory without being excessive and when the temperture steadily rises as summer approaches. In contrast, high temperature and lack of moisture are adverse to vector breeding. Elsewhere, potato spotted wilt is more prevalent in dry rather than in wet seasons.

Other Hosts

TSWV infects dicots and monocots of more than 30 families, annuals and perennials, and many crop, weed, and ornamental plants. Tomato and tobacco are much more severely damaged than is potato.

Petunia hybrida is a local lesion host. (Lesions appear in two to four days on intact or detached leaves.) It is usually not systemically infected.

Phaseolus vulgaris (Manteiga and related cultivars) exhibits chlorotic local lesions in three to six days (Fig. 93E).

Nicotiana tabacum (Samsun NN, Turkish NN, Blue Pryor), *N. clevelandii*, *N. glutinosa*, and *N. rustica* show local lesions, systemic necrosis, and leaf deformation. They are used for local lesion assay and virus purification.

Tropaeolum majus, *Cucumis sativus*, and *Vinca rosea* are good diagnostic species.

T. majus and *Gomphrena globosa* are used for maintaining cultures.

Resistance

Some field resistance is present in Kathadin and Snowflake, in some of their hybrids, and in Brownell, Brown's, and Epicure. This level of field resistance breaks down easily under severe exposures. Potato is more susceptible to infection during the growing phase than it is after flowering, when plants become increasingly resistant.

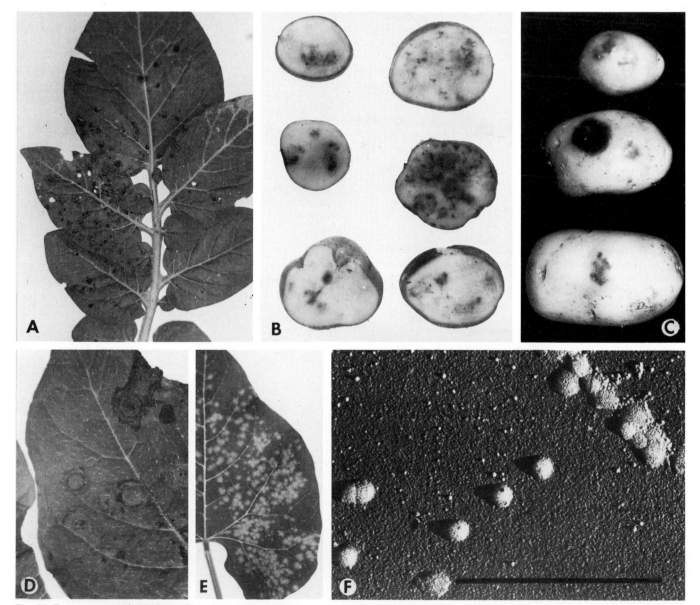

Fig. 93. Tomato spotted wilt: **A** and **D**, lesions on potato leaves; **B** and **C**, symptoms within and on tubers; **E**, local lesions on bean leaves; **F**, purified virus. Bar represents 1 μm. (A–E, Courtesy A. S. Costa; F, courtesy E. W. Kitajima)

Control

1) Avoid locations in which severe outbreaks have occurred frequently, and because of virus reservoirs in ornamental and vegetable plantings, avoid locations near residences.

2) Select higher elevations when possible.

3) Early cultivars and early plantings may escape infection.

4) Insecticides (foliage sprays or systemic granulated compounds) in the planting or on border catch crops have been used with varying effectiveness in various crops.

Selected References

BEST, R. J. 1968. Tomato spotted wilt virus. Adv. Virus Res. 13:65–146.

COSTA, A. S., and J. KIEHL. 1938. Una molestia da batatinha "necrose do topo" causada pelo virus de vira-cabeça. J. Agron., Piracicaba 1:193–202.

IE, T. S. 1970. Tomato spotted wilt virus. No. 39 in: Descriptions of Plant Viruses. Commonw. Mycol. Inst., Assoc. Appl. Biol., Kew, Surrey, England.

KITAJIMA, E. W. 1965. Electron microscopy of vira-cabeça (Brazilian spotted wilt virus) within the host cell. Virology 26:89–99.

MAGGEE, C. J. 1936. Spotted wilt disease of lettuce and potatoes. Agric. Gaz. N.S.W. 47:99–100.

NORRIS, D. O. 1951. Spotted wilt of potato. Aust. J. Agric. Res. 2:221–260.

TAS, P. W. I., M. L. BOERJAN and D. PETERS. 1977. Purification and serological analysis of tomato spotted wilt virus. Neth. J. Plant Path. 83:61–72.

(Prepared by A. S. Costa and W. J. Hooker)

Potato Spindle Tuber Viroid

Reliable reports of potato spindle tuber viroid (PSTV) causing spindle tuber in commercial plantings have been made from the United States, Canada, and the USSR. The viroid also causes a tomato disease in South Africa.

Symptoms

Vine symptoms are seldom evident before blossom time. Stems and blossom pedicels are slender, longer than normal, and remain upright. Leaflets, slightly small and with fluted margins, tend to curve inward and overlap the terminal leaflet. Angles between stems and petioles are more acute than normal.

Leaves near the ground are noticeably shorter and upright, contrasting with healthy leaves, which rest on the ground (Fig. 94). As the season advances, diseased plants are restricted in growth and become harder to identify because of intertwining with neighboring healthy plants. The severe strain (unmottled curly dwarf) causes enhanced symptoms, twisting of leaflets, and rugosity of leaf surfaces. Under some light conditions, plants with spindle tuber show a dull leaf surface, less reflective of light than normal leaf surfaces.

Tubers are elongated, round in cross section, and tend in some cultivars to have pointed ends. The pointed stem end is more characteristic than elongation; the round cross section, in contrast to the flat cross section of normal tubers, is a diagnostic feature (Fig. 95A and B). Tuber symptoms become more marked as the season advances. Russet skins become smooth; red skins become pink; and purple skins turn a lighter lavender. Eyes appear to be more numerous; they have characteristic indentation with heavy brows. Necrotic spots often appear around lenticels. Surface cracking, usually parallel to the long axis, is frequent. In some cultivars, knobs and swellings appear and tubers are severely misshapen. Extensive necrotic tissue may also occur in the flesh. Not all tubers from diseased plants show all or even any of these symptoms, however, and some tubers from healthy plants resemble spindle tubers. Attempting to reduce spindle tuber by sorting out diseased tubers is not effective.

In potato, mild strains with indistinct symptoms outnumber severe strains by a ratio of 10:1 and cause yield losses of 15–25%, whereas severe strains with distinct symptoms cause a 65% yield loss.

Causal Agent

PSTV is a viroid, an extremely small RNA molecule with molecular weight of 100,000–125,000. It is circular and has no protein coat. Thermal inactivation is 75–80° C in 10 min and, in phenol-treated preparation, 90–100° C. Other plant viroids cause similar symptoms in tomato. Polyacrylamide gel electrophoresis

Fig. 95. Potato spindle tuber viroid: **A,** symptomless tuber; **B,** infected tuber; **C,** local lesions on *Scopolia sinensis* (right half of leaf was inoculated). **D,** Tomato plants 56 days old: **left,** healthy; **center,** inoculated with potato spindle tuber viroid at 14 days; **right,** inoculated in cotyledon stage. (Courtesy R. P. Singh)

Fig. 94. Spindle tuber viroid: **A,** early symptoms; **B,** symptomless plants.

shows PSTV nucleic acid produced by all strains of PSTV as a band additional to those in healthy plants.

Epidemiology

Transmission is largely mechanical, principally by man himself and to a lesser extent by chewing insects. Evidence that sucking insects are involved is inconclusive. This is one of the few potato diseases transmitted readily by pollen and true seed.

Other Hosts

About two to three weeks after inoculation of *Lycopersicon esculentum* with severe strains, new leaflets show marked rugosity, epinasty, and down-curling (Fig. 95D). Internodes become shorter, forming a rosette or bunchy top. Later veinal necrosis can become very severe. With mild strains, symptoms are slight epinasty and stunting. Bunchy top is increased in continuous light of 1,000 FC or more. Mild strains temporarily cross-protect against later inoculation with severe strains, and the reaction can be used to demonstrate the presence of mild strains in otherwise symptomless plants.

Scopolia sinensis responds with dark brown, necrotic local lesions in two or three weeks and later with systemic necrosis (Fig. 95C). Severe strains cause symptoms earlier than do mild strains. Optimum conditions are manganese-rich soil nutrition, 18–23°C incubation in not over 300–400 FC of light, preinoculation shading, and inoculum in $0.05 M$ K_2HPO_4, pH 9.0, buffer. Certain insecticides severely impair local lesion formation.

Many plant species belonging to most genera of the Solanaceae are symptomlessly infected. PSTV also infects members of the Amaranthaceae, Boraginaceae, Campanulaceae, Caryophyllaceae, Compositae, Convolvulaceae, Dipsaceae, Sapindaceae, Scrophulariaceae, and Valeriaceae.

Control

1) Use seed tubers known to be free from PSTV, such as government-inspected certified seed.

2) Avoid mechanical transmission by planting whole, rather than cut, seed, and avoid leaf contact by equipment in field operations.

3) Decontaminate knives and other equipment by dipping them in or washing them with sodium hypochlorite (0.25%) or calcium hypochlorite (1.0%).

4) Roguing of diseased plants in seed fields is ineffective because of indistinct plant symptoms.

5) Plant seed fields with whole (uncut) tubers and wide spacing for large tubers or by the tuber unit method; the latter aids in disease identification but introduces the danger of transmission during cutting.

Selected References

DIENER, T. O., and A HADIDI. 1977. Viroids. Comp. Virol. 11:285–337.

DIENER, T. O., and W. B. RAYMER. 1971. Potato spindle tuber "virus." No. 66 in: Descriptions of Plant Viruses. Commonw. Mycol. Inst., Assoc. Appl. Biol., Kew, Surrey, England.

FERNOW, K. H. 1967. Tomato as a test plant for detecting mild strains of potato spindle tuber virus. Phytopathology 57:1347–1352.

McCLEAN, A. P. D. 1948. Bunchy-top disease of the tomato: Additional host plants, and the transmission of the virus through the seed of infected plants. S. Afr. Dep. Agric. Sci. Bull. 256. 28 pp.

MORRIS, T. J., and N. S. WRIGHT. 1975. Detection on polyacrylamide gel of a diagnostic nucleic acid from tissue infected with potato spindle tuber viroid. Am. Potato J. 52:57–63.

SINGH, R. P. 1973. Experimental host range of the potato spindle tuber "virus." Am. Potato J. 48:262–267.

SINGH, R. P., R. E. FINNIE, and R. H. BAGNALL. 1971. Losses due to the potato spindle tuber virus. Am. Potato J. 48:262–267.

YANG, T. C., and W. J. HOOKER. 1977. Albinism of potato spindle tuber viroid-infected Rutgers tomato in continuous light. Am. Potato J. 54:519–530.

(Prepared by R. P. Singh and K. H. Fernow)

Sugar Beet Curly Top Virus

Curly top, caused by the sugar beet curly top virus (BCTV), moves slowly through the potato plant and is apparently now of little importance.

Current season symptoms consist of dwarfing, yellowing, elongation, and upward rolling of the midrib of terminal leaflets. Leaflets near the growing point have marginal yellowing, elongation, cupping, rolling, twisting, bulging, and roughness (Fig. 96). Tuber-perpetuated symptoms are extremely variable. Tubers from infected plants may produce healthy appearing plants, sprouts delayed in emergence or failing to emerge, or sprouts having extremely short shoots with unmottled leaves ranging from dark to light green and yellow. The green dwarf symptom consists of shoots delayed in emergence with leaves failing to unfold completely and remaining stiff and erect. As stems elongate, they stay stiff and erect, with the growing point pinched together.

Symptoms are remarkably similar to certain symptoms of the mycoplasma yellows disease and therefore may have escaped identification.

Transmission in nature is by the leafhopper, *Circulifer (Neoliturus) tenellus*, in which the virus is circulative.

Selected References

BENNETT, C. W. 1971. The curly top disease of sugarbeet and other plants. Monogr. 7. Am. Phytopathol. Soc., St. Paul, MN. 81 pp.

GARDNER, W. S. 1954. Curly top of the potato in Utah and its possible relationship to "haywire." Plant Dis. Rep. 38:323–325.

GIDDINGS, N. J. 1954. Some studies of curly top on potatoes. Phytopathology 44:125–128.

STOKER, G. L., and O. S. CANNON. 1962. Current-season and tuber-perpetuated symptoms of potato curly top. Plant Dis. Rep. 46:176–180.

(Prepared by W. J. Hooker)

Fig. 96. Sugar beet curly top virus, showing leaf rolling and dwarfing. (Courtesy N. J. Giddings, from Bennett, C. W., 1971)

Mycoplasmas

The "yellows" types of disease, characteristic of mycoplasma infections that have been studied in some detail possess rather broad host ranges. Some have been reported to infect potato in the field or have been transmitted to potato experimentally. Unfortunately, disease symptoms in the potato itself are hardly of diagnostic value except to distinguish two large groups of agents, those causing the aster yellows and the witches' broom types of disease.

Mycoplasma-like organisms (MLO), formerly considered viruses with somewhat unusual characteristics, were first demonstrated in plants in 1967 by electron microscopy. Actual proof that MLO cause disease in potato is needed by Koch's postulates, although the presence of MLO in plant tissue constitutes the best evidence so far available. More accurate methods of differentiating between these pathogens are urgently needed, particularly because potato is a relatively incompatible host.

None of these diseases is contact-transmissible. Grafting is often used for experimental transmission. All of the pathogens rely on leafhoppers for transmission and dispersal, and their occurrence and distribution is determined by leafhopper activity. Variations in symptomatology, host range, or vector relations of different source materials of each disease suggest that strains of these pathogens occur.

Symptoms and MLO may be suppressed by antibiotics of the tetracycline group, and individual plants may be cured by heat therapy.

Aster Yellows and Stolbur

Aster yellows and its allied diseases occur worldwide. Stolbur is found in Europe, tomato big bud in Australia, purple top roll in the Indian peninsula of Asia, and parastolbur and metastolbur in Europe. In the western hemisphere, aster yellows has been variously called purple top wilt, yellow top, bunch top, purple dwarf, apical leafroll, haywire, latebreaking virus, blue stem, and moron.

Symptoms

Upper leaflets roll and develop purple or yellow pigmentation (Plate 77). Aerial tubers are common, and occasionally some proliferation of axillary buds occurs (Plate 78). Often only a single stem in a hill is affected. Plants are often stunted and may die prematurely (Fig. 97A and B). Under field conditions, lower stems frequently develop cortical necrosis, shredding of tissue, and vascular discoloration.

At harvest, an affected hill usually has some normally mature and some immature tubers. In stolbur, flaccid (gummy) tubers

Fig. 97. Aster yellows mycoplasma in potato: **A,** primary symptoms; **B,** advanced wilting of stolbur in left stem with stem at right unaffected; **C,** aster yellows mycoplasma in infected tissue of aster (electron microscope photograph, bar represents 1 μm. (A and C, Courtesy N. S. Wright; B, courtesy V. Valenta)

may form.

When the causal agents survive storage in seed tubers, plants in the second year develop symptoms. In the more usual situation, in which causal agents fail to survive in seed tubers, plants in the second year may be normal, lack diagnostic symptoms, be smaller than normal, or fail to emerge. Hair sprouts frequently develop on tubers from infected plants (Plate 79). Plants with stolbur may bear simplified "round" leaves, but this symptom disappears in subsequent years.

Causal Organisms

Mycoplasmalike organisms (MLO) occur in phloem sieve cells and occasionally in phloem parenchyma cells of infected plants (Fig. 97C). They are pleomorphic, lack a cell wall, and are bounded by a unit membrane. MLO vary in diameter from 50 to 1,000 nm. The larger, more prevalent forms are roughly spherical and contain a central fibrillar network of strands, presumably DNA, and a peripheral area of ribosomelike granules. The presence of elongate forms and small, dense "elementary bodies" suggests propagation by binary fission, budding, or fragmentation.

Epidemiology

The principal leafhopper vector of aster yellows, *Macrosteles fascifrons*, overwinters on weeds, grasses, and small grains. The favorite host of the main vector of stolbur, *Hyalesthes obsoletus*, is perennial bindweed, *Convolvulus arvensis*, which simultaneously acts as the main inoculum reservoir plant.

All pathogens are transmitted by and propagative in leafhoppers. Potato is not a favored host and usually escapes extensive disease incidence. Stolbur is an exception that, in certain European regions, may become very destructive. Leafhopper vectors do not complete their life cycles on potato. Neither nymphs nor adults can acquire the pathogen from potato. Transmission to potato occurs only when the leafhopper has fed on some other infected hosts two to three weeks before feeding on potato.

Other Hosts

Many vegetable, ornamental, field crop plants, and weeds are among the 350 species from at least 54 families susceptible to aster yellows.

Witches' Broom

Witches' broom, also called northern stolbur and dwarf shrub virosis, has been reported from North America, Europe, and Asia but is of minor economic importance. Several distinct pathogens or strains of a single pathogen may be involved in different geographical areas.

Symptoms

Symptoms are generally similar. Plants have many axillary and basal branches with simple "round" leaves, which may be chlorotic (Fig. 98A). This disease is distinct from wilding, a somatic aberration. Current symptoms appear late, if at all, and consist of an upright habit of growth, rolling of leaflets, and some chlorotic or anthocyanin discoloration in apical leaves. Most strains cause no flower symptoms, but witches' broom strains of central and eastern Europe are characterized by virescence, phyllody, and associated flower symptoms (Fig. 98B).

Mycoplasmas of witches' broom are tuber-perpetuated, in contrast to those of aster yellows, which usually are not. Affected hills usually produce tubers that appear normal during the year of infection but give rise to plants with advanced symptoms the subsequent year. Tubers from infected plants frequently produce hair sprouts. Such plants have many small tubers and an abnormally high number of sprouts with simplified, small leaves. Occasionally these symptoms follow

early infections during the current season. Infected tubers usually lack dormancy.

Causal Organism

Pathogens are mycoplasmalike organisms, such as those that cause aster yellows.

Epidemiology

The pathogens are transmitted by leafhopper vectors, *Ophiola (Scleroracus) flavopictus, S. dasidus* (Fig. 98C), and *S. balli*, in which they are propagative. In most areas, the vectors are not yet known, nor are the natural host ranges of the pathogens.

Leafhoppers are unable to acquire the pathogen from potato. Spread to potato occurs when a vector feeds on other infected hosts several days before feeding on potato. Witches' broom in perennial legumes is usually more serious than in potato because the former are preferred hosts by the vector and inoculum is usually available within the crop.

Other Hosts

Lycopersicon esculentum, Cyphomandra betacea, and *Nicotiana tabacum* have been used to distinguish between strains.

Medicago sativa, Melilotus alba, or *Lotus corniculatus* serve as natural reservoirs and may be used as indicator plants.

Trifolium repens is also susceptible.

Control

1) Generally these diseases are of minor importance and do not justify elaborate control measures.

2) Control leafhopper migrations into potato fields by

Fig. 98. Witches' broom: **A,** in potato plant; **B,** current season, medium intensity flour phyllody of Czechoslovak type II; **C,** *Scleroracus dasidus* female, one of the leafhopper vectors. (A, Courtesy N. S. Wright; B, courtesy V. Valenta: C, courtesy J. Raine)

modifying cultural practices and by using insecticides.

3) Spread from potato to potato is believed not to occur.

4) Heat treatment may be useful on individual plants.

5) Remission of symptoms often follows treatment with tetracyclines.

Selected References

BRCÁK, J., O. KRÁLÍK, J. LIMBERK, and M. ULRYCHOVÁ. 1969. Mycoplasm-like bodies in plants infected with potato witches' broom disease and the response of plants to tetracycline treatment. Biol. Plant. 11:470–476.

CHAPMAN, R. K., moderator. 1973. Symposium: Aster yellows. Proc. N. Cent. Branch, Entomol. Soc. Am. 28:38–99.

DOI, Y., M. TERAMAKA, K. YORA, and H. ASUYAMA. 1967. Mycoplasma or PLT group-like microorganisms found in the phloem elements of plants infected with mulberry dwarf, potato witches' broom, aster yellows, or Paulownia witches' broom. Ann. Phytopathol. Soc. Japan 33:259–266.

FUKUSHI, T., E. SHIKATA, H. SHIODA, E. SEKIYAMA, L. TANAKA, N. OSHIMA, and Y. NISHIA. 1955. Insect transmission of potato witches' broom in Japan. Proc. Jpn. Acad. (Nihon Gakushuiin) 31:234–236.

NAGAICH, B. B., B. K. PURI, R. C. SINHA, M. K. DHINGRA, and V. P. BLARDWAJ. 1974. Mycoplasma-like organisms in plants affected with purple top-roll, marginal flavescence and witches' broom diseases of potatoes. Phytopathol. Z. 81:273–279.

RAINE, J. 1967. Leafhopper transmission of witches' broom and clover phyllody viruses from British Columbia to clover, alfalfa, and potato. Can. J. Bot. 45:441–445.

SEMANCIK, J. S., and J. PETERSON. 1971. Association of a mycoplasma with haywire disorder of potatoes. Phytopathology 61:1316–1317.

VALENTA, V. 1957. Potato witches' broom virus in Czechoslovakia. Pages 246–250 in: F. Quak, J. Dijkstra, A. B. R. Beemster, and J. P. H. Van der Want, eds. Proc. Third Conf. Potato Virus Dis. 24–28 June, 1957. H. Veenman and Zonen, Lisse-Wageningen, The Netherlands. 282 pp.

VALENTA, V. 1969. Vergleich eines aus Niedersachsen stammenden Kartoffelhexenbesen-Virus mit anderen aus Europa bekannten Viren dieser Gruppe, Zentralbl. Bakteriol. Parasiteuk. Infektionskr. Hyg. Abt. 2. 123:352–357.

VALENTA, V., M. MUSIL, and S. MIŜIGA. 1961. Investigations on European yellows-type viruses. I. The stolbur virus. Phytopathol. Z. 42:1–38.

VALENTA, V., and M. MUSIL. 1963. Investigations on European yellows-type viruses. II. The clover dwarf and parastolbur viruses. Phytopathol. Z. 47:38–65.

WRIGHT, N. S. 1957. Potato witches' broom in North America. Pages 239–245 in: F. Quak, J. Dijkstra, A. B. R. Beemster, and J. P. H. Van der Want, eds. Proc. Third Conf. on Potato Virus Dis, 24–28 June, 1957. H. Veenman and Zonen, Lisse-Wageningen, The Netherlands. 282 pp.

(Prepared by N. S. Wright, J. Raine, and V. Valenta)

Insect Toxins

Psyllid Yellows

This disorder results from insect feeding, and no infectious microorganism is involved.

On vines, young leaves, which are often red or purple, become erect and have cupped basal portions. Nodes of young stems become enlarged; the axillary angle between stems and petioles is larger than usual; and aerial tubers or stocky shoots, frequently rosetted, form at the leaf axils. Plants have a pyramidal shape and are generally yellow or bronzed. In advanced stages, older leaves roll upward, become yellow with necrotic areas, and break down rapidly. The tip leaves form a rosette (Plate 80). Plants do not generally wilt even under extreme drought.

Border parenchyma surrounding the phloem is first affected, and later tissue breakdown extends laterally, causing phloem necrosis. Abnormally large deposits of starch develop in the cortex and pith.

Few if any tubers are set on plants attacked in early development. When older plants are affected, stolon tips may produce sprouts and secondary shoots that emerge from the soil as small plants. Masses of small tubers may form on secondary stolon branches, or successive tubers may form on a stolon in a chain or beadlike arrangement.

Abnormally small tubers sprout without a dormant period. Similarly, dormancy may be lacking in tubers that appear normal. Because of the many small tubers, the number of marketable tubers is greatly reduced.

The disorder is not tuber-transmitted. Artificial methods of transmission from plant to plant have not been successful.

Tubers from affected plants produce plants that appear normal except for a slightly spindly or weakened appearance. Seed tubers from affected fields are not of satisfactory quality.

The disease results from toxic substances introduced during feeding of nymphs of *Paratrioza cockerelli*, the tomato or potato psyllid, also known as the jumping plant louse. As few as three to five nymphs may cause early symptoms, but higher populations (15–30 per plant) are required for full symptom expression. Adults up to 1,000 per plant do not induce symptoms. Progress in symptom development is stopped abruptly when nymphs are removed, and affected plants frequently recover.

During the 1930s, the disease caused considerable damage in the United States west of the Missouri River and extending from New Mexico into Canada. In certain areas it was the most destructive disease. Symptoms are most severe in high intensity sunlight, at high field temperatures, and during drought.

Similarity between psyllid yellows, the mycoplasma disorder haywire, leaf rolls of various types, and Rhizoctonia has caused some confusion in diagnosis.

Selected References

EYERS, J. R. 1937. Physiology of psyllid yellows of potatoes. J. Econ. Entomol. 30:891–898.

RICHARDS, B. L., and H. L. BLOOD. 1933. Psyllid yellows of the potato. J. Agric. Res. 46:189–216.

SCHAAL, L. A. 1938. Some factors affecting the symptoms of the psyllid yellows disease of potatoes. Am. Potato J. 15:193–206.

(Prepared by W. J. Hooker)

Nematodes

Nematodes pathogenic to potatoes (Table II) occur in all climates and cause serious crop losses, but much of this damage is unrecognized or attributed to other causes. Because nematodes attack roots and tubers, no diagnostic symptoms appear on aboveground parts of the plant except for unthrifty top growth resulting from poor root systems. Low densities in the soil cause no top symptoms but may reduce tuber yields. As the world population increases, soil suitable for potato culture will become more scarce. Consequently, potatoes will be grown more frequently on the best potato land, and because monoculture encourages nematode population increase, nematode damage to potatoes will increase dramatically.

Confining nematode populations to areas where they already exist by restricting movement of infected seed tubers and plants may be the most effective way of preventing loss of productive land. Care in purchase of seed and prevention of shipment of infected seed into nematode-free areas cannot be over emphazied.

Selected Reference

WINSLOW, R. D., and R. J. WILLIS. 1972. Nematode diseases of potatoes. Pages 17–48 in: J. M. Webster, ed. Economic Nematology. Academic Press, London and New York. 563 pp.

Potato Cyst Nematodes

Cyst nematodes, *Globodera (Heterodera)* spp., also known as golden nematodes or potato root eelworms, are present in most countries of northern and central Europe and to a lesser extent in southern Europe, Newfoundland, British Columbia, Greece, Israel, Tunisia, South Africa, New Zealand, and Japan. Reports from Asia are incomplete.

Both *G. rostochiensis* and *G. pallida* occur in South America and Europe. Only *G. pallida* has been identified in Colombia, Ecuador, and in most of Peru. In southern Peru, Boliva, and Argentina, *G. pallida* and *G. rostochiensis* occur together, and in Chile, Venezuela, Central America, and Mexico, only *G. rostochiensis* has been found. Both species occur in the central and western European countries. In the south and east of Europe, only *G. rostochiensis* is found. In the United Kingdom, certain potato-growing areas have predominantly either *G. rostochiensis* or *G. pallida*. Only *G. rostochiensis* occurs in the United States. In many countries and areas of countries, species determinations have not been made.

Symptoms

Cyst nematodes do not cause specific aboveground symptoms of diagnostic value, but root injury causes infected plants to seem to be under stress from water or mineral deficiency. Foliage is pale, and severe wilting occurs under drought. Large nematode populations cause stunting, early senescence, and often proliferation of lateral roots. At flowering, minute

---

TABLE II. Nematode Pests of the Cultivated Potato

Scientific and Common Names	Distribution[a] H	C	S	T	Transmission by Tubers[b]
Belonolaimus longicaudatus (sting nematode)	h			t	−
Ditylenchus destructor (potato-rot nematode)			s	T	+
Ditylenchus dipsaci (stem and bulb nematode)		c		t	+
Helicotylenchus spp. (spiral nematodes)	h	c	s	t	+
Globodera spp. (round cyst nematodes)					
G. pallida—potato cyst nematode (white immature females)		C		T[c]	+
G. rostochiensis—potato cyst nematode (golden immature females)		C		T[c]	+
Hexatylus vigissi				t	+
Longidorus maximus (needle nematode)				t[d]	−
Meloidogyne spp. (root-knot nematodes)					
M. acronea				t	+
M. africaná	h			t	+
M. arenaria	h	c	s	T	+
M. hapla		C	s	T	+
M. incognita	H	c	S	T	+
M. javanica	H	c	S	T	+
M. thamesi				t	+
Meloinema sp.			s	t	+
Nacobbus aberrans (false root-knot nematode)	h	C		T	+
Neotylenchus abulbosus				t	+
Paratylenchus spp. (pin nematodes)	h	c	s	t	−+
Pratylenchus spp. (root-lesion nematodes)					
P. andinus		c			+
P. brachyurus			s	t	+
P. crenatus				t	+
P. coffeae	h		s	t	+
P. minyus		c	s	t	+
P. penetrans	h	C	s	T	+
P. pratensis		c	s	t	+
P. scribneri	h	c	s	t	+
P. thornei		c	s	t	+
Rotylenchulus spp. (reniform nematodes)	h	c		t	+
Rotylenchus spp. (spiral nematodes)	h	c		t	+
Trichodorus spp., *Paratrichodorus* spp. (stubby root nematodes)					
T. allius				t[d]	−
P. christiei		c		t[d]	−
P. pachydermus		c		t[d]	−
T. primitivus		c		t[d]	−
P. teres				t[d]	
Tylenchorhynchus spp. (stunt nematodes)					
T. claytoni	h	c	s	t	−
T. dubius		c	s	t	−
Xiphinema spp. (dagger nematodes)	h	c	s	t[d]	−

[a] H = hot tropical, C = cool tropical, S = subtropical, T = temperate zone. When capitalized, as shown, the judged relative importance is greater than when small letters are used. All attack potato and additional members of the Solanaceae as well as plants outside the Solanaceae.
[b] None are known to be transmitted by true botanical seed.
[c] Limited to the Solanaceae.
[d] Plant virus vectors.

immature females in the white or yellow stage erupt through the root epidermis. Yield losses vary according to nematode densities, and complete economic crop failure can result when densities are high. Potato cyst nematodes may increase incidence of Verticillium wilt and bacterial wilt (brown rot).

Causal Organisms

G. rostochiensis and *G. pallida* become round cysts upon maturity. Cysts are light to dark brown with an irregular pattern of subsurface punctuations over most of the body.

Cysts of *G. rostochiensis* differ from those of *G. pallida* by a greater average anal-vulval distance, 60 compared with 44 μm, and a greater average number of cuticular ridges between anus and vulva, 21.6 compared with 12.2. *G. rostochiensis* females develop through a golden yellow phase before turning brown, hence the common name, golden nematode; *G. pallida* females are white or cream before turning brown (Fig. 99A, Plates 81 and 82). Lengths of larval body, stylet, and tail are usually longer in *G. pallida* than in *G. rostochiensis*.

Races are differentiated by ability to multiply on resistant cultivars of *S. tuberosum* ssp. *andigena*, *S. multidissectum*, *S. vernei*, or *S. kurtzianum*.

Disease Cycle

In the spring, over 50% of the second stage larvae inside eggs within a cyst are stimulated to hatch. They enter the host plant roots, feed, and develop through a series of three molts. The females enlarge and rupture the root tissue but remain attached to the root by their heads and protruding necks, which stay inserted in root tissue. Mature wormlike males leave the root and mate with the females. Fertilized females increase in size to become subspherical. Mature females measure between 0.5 and 0.8 mm in length and greatly vary in size, probably due to the type of host and amount of nutrition during their development. Eggs are produced and retained within the female. The female cuticle darkens and hardens, becoming the cyst, which may contain as many as 500 eggs (Figs. 99B and 100A). Cysts remain in the soil when the crop is harvested.

Multiplication rate and sex ratio are influenced by population density of the nematode and host crop. An ample food supply favors a multiplication rate up to 60-fold. When food is limited and the population is large (100 eggs per gram of soil) nematode density may decline.

Histopathology

Globodera spp. are stimulated to hatch by exudates from plant roots. Second stage larvae usually enter the root hair zone. As larvae move through cortex cells of potato roots, feeding may cause some limited necrosis in susceptible cultivars.

The female feeds near the vascular cylinder, resulting in multinucleate units called syncytia (giant cells) near the nematode's head. Syncytia are formed by incorporation of adjacent cells following cell-wall dissolution, which begins in the cortex. Cell walls of a column of cells toward the vascular tissue are then dissolved. In the vascular cylinder, syncytia are limited by lignified xylem, so incorporation of new cells proceeds parallel to the root axis (Fig. 100B).

Syncytia may be formed in the cortex, endodermis, pericycle, and parenchyma of the central vascular strand. Cytoplasm of syncytia becomes dense and granular in structure.

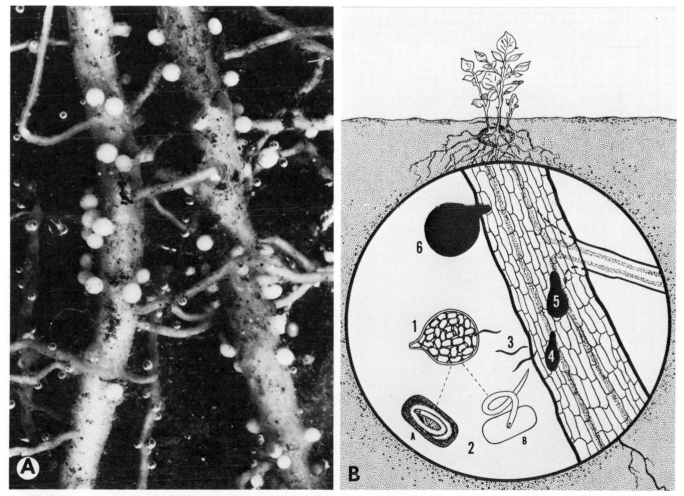

Fig. 99. Potato cyst nematode (*Globodera rostochiensis*). **A,** immature swollen females attached to potato roots. **B,** diagrammatic life cycle of potato cyst nematodes: **1,** cyst showing enclosed eggs; **2,** enlarged egg showing enclosed, coiled larva; **3,** larvae entering root; **4** and **5,** swollen females feeding in root; **6,** mature female breaking through root surface. (Courtesy W. F. Mai, B. B. Brodie, and M. B. Harrison)

Syncytia usually are elongate, with ends merging with normal tissue, and each syncytium is generally associated with one larva. When multiple infections occur within a small area of root tissue, syncytia may coalesce. Nuclear hypertrophy is followed by decrease in size and number of plastids, breakdown of chondriosomes (mitochondira), polyploidy of nuclei, and nuclear disintegration.

Ingrowths or protuberances develop next to xylem vessels; "boundary formations" and microtubules are associated with the ends of these protuberances. They serve to increase the surface area of the syncytial cell wall relative to its volume and to allow for increased flow of solutes across the plasmalemma. The cell wall becomes up to 10 times its normal thickness.

Epidemiology

Although populations of cyst nematodes do not increase as rapidly as do fungal and bacterial pathogens of potatoes, once well established in a potato-growing area, they are, with present technology, impossible to eradicate. The environmental conditons providing successful commercial potato production also provide optimum conditions for their multiplication and survival. Potato cyst nematodes flourish where soil temperatures are cool. Although they have been found in tropical and warmer temperate climates, they do not generally become established and are of lesser economic importance than in cool climates. Larvae become active at 10° C, and maximum invasion of roots occurs at 16° C. Soil temperatures of 26° C for prolonged periods of time reduce development and limit reproduction.

Cyst nematodes develop well in soils suited for survival and movement of wormlike stages, such as medium to heavy clay soils and well-drained and aerated sands, silts, and peat soils with a moisture content of 50–75% of water capacity. Soil pH values that are tolerable to the potato plant can apparently be tolerated by the nematodes. Nutritional status of the soil appears to have little or no effect on nematodes other than that caused by crop performance.

Encysted eggs withstand desiccation and can remain viable 20 years or more in soil under severe environmental extremes. Moving infested soil such as that clinging to equipment, seed, or storage containers is the most important means of local and long distance spread. Planting contaminated tubers provides ideal conditions for spread and is thought to be a primary factor in nematode dissemination throughout the world. Birds are not considered important in long-distance spread.

Other Hosts

These include tomato, eggplant, and a number of Solanaceous weeds.

Resistance

Resistant cultivars and nonhost crops cause an average of 95 and 50% reduction in populations, respectively. Excellent sources of resistance to G. rostochiensis (race R_1A) are available in commercial varieties in Europe and North America. Good resistance has been found to some, but not all, races of G. pallida. Resistance to G. pallida (race P_4A) is available in some newer Dutch varieties.

Control

1) Restrict shipments of seed tubers and plants of other types from infested areas.

2) Except for high dosages of soil fumigants, chemical treatments usually reduce densities only slightly, if at all. Although some organic phosphate and carbamate nematicides provide good protection against infection by active larvae, nematode density in treated soil usually remains the same or slightly increases during growth of a potato crop.

3) Crop rotation has been widely used but is often uneconomical because of the length of rotation required. When nematode densities are high, rotation with potatoes grown once in five years is necessary to assure profitable potato yields. Resistant potato cultivars in rotation with susceptible cultivars and nonhosts considerably reduce the required length of rotation.

4) Combining different control measures is necessary for keeping populations below damaging levels and for preventing establishment of nematodes in new areas. Key components of nematode management are: extensive surveys to determine distribution of cyst nematodes, soil fumigants to reduce numbers of nematodes in the soil, resistant cultivars to prevent density increase, carbamate nematicides to suppress density increases, prohibition of potato seed production in known infested or exposed land, and regulation of reuseable containers and movement of farm machinery, top soil, and plant material.

Selected References

CHITWOOD, B. G., and E. M. BUHRER. 1946. The life history of the golden nematode of potatoes, *Heterodera rostochiensis* Wollenweber, under Long Island, New York, conditions. Phytopathology 36:180–189.

ENDO, B. Y. 1971. Nematode-induced syncytia (giant cells). Pages 91–117 in: B. M. Zuckerman, W. F. Mai, and R. A. Rohde, eds. Plant Parasitic Nematodes, Vol. 2. Academic Press, New York. 347 pp.

EVANS, K., J. FRANCO, and M. M. de SCURRAH. 1975. Distribution of species of potato cyst-nematodes in South America. Nematologica 21:365–369.

HOOPES, R. W. 1977. The internal response of several resistant and susceptible potato clones to invasion by the potato cyst nematode *Heterodera rostochiensis* Wollenweber. MS thesis, Cornell University. 61 pp.

MULVEY, R. H., and A. R. STONE. 1976. Description of *Punctodera matadorensis* n. gen., n. sp. (Nematoda: Heteroderidae) from Saskatchewan with lists of species and generic diagnosis of Globodera (n. rank), Heterodera and Sarisodera. Can. J. Zool. 54:772–785.

SPEARS, J. F. 1968. The Golden Nematode Handbook: Survey,

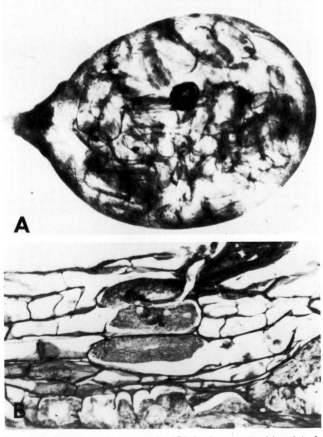

Fig. 100. Potato cyst nematode (*Globodera rostochiensis*): **A,** mature cyst with enclosed eggs; **B,** section of potato root showing syncytia. (Courtesy W. F. Mai, B. B. Brodie, and M. B. Harrison)

Laboratory, Control, and Quarantine Procedures. U. S. Dept. Agric. Handbook 353. 81 pp.

STONE, A. R. 1972. *Heterodera pallida* n. sp. (Nematoda: Heteroderidae), a second species of potato cyst nematode. Nematologica 18:591–606.

Root-Knot Nematodes

One or more species of *Meloidogyne* are known to attack almost all major crop plants and many weeds species. Vegetable crops, including potatoes, are extensively damaged, with potato losses reaching 25% or more. Although species differ in their ability to attack certain vegetable crops, no vegetables go unharmed.

Root-knot nematodes are worldwide in distibution but are limited in specific areas by temperature and cropping practices. The *M. incognita* group is, perhaps, the most widely distributed. *M. hapla* is the dominant species on potato in Europe and North America followed by *M. incognita* and *M. incognita acrita*. In Africa and Asia, *M. javanica* and *M. incognita* are dominant, followed by *M. incognita acrita* and *M. hapla*, the latter being found in Japan. *M. incognita*, *M. incognita acrita*, *M. javanica*, and *M. hapla* attack potatoes in South America. *M. arenaria* has been found in potatoes on most continents.

Symptoms

Aboveground symptoms are not diagnostic. Depending upon nematode density, infected plants may show varying degrees of stunting and a tendency to wilt under moisture stress.

Knots or galls of varying sizes and shapes are present on the roots (Fig. 101A). When nematode densities are high and environmental conditions favorable, tubers are infected and display galls that give them a warty appearance (Fig. 101B). Galls, containing white, pear-shaped, mature female nematodes, range from an almost spherical shape (in *M. arenaria*) to a very rough and irregular appearance (in *M. hapla*). Individual gall size depends upon nematode density and species, root size, temperature, and other environmental factors. In addition to galling, *M. hapla* causes initiation of extensive lateral root formation.

Disease Cycle of Causal Organism

The disease cycle of root-knot nematodes on potato is similar to that on other crops and plants. The first molt occurs within the egg; the second-stage larva emerges from the egg and invades the host root near the tip. Larvae migrate through the root to the vascular tissue, where they become stationary. Feeding injury and glandular secretion by larvae cause host cells surrounding the nematode head to undergo cell division and cell enlargement. Interaction of nematode and host causes development of multinucleate giant cells, from which the nematode obtains its food. After feeding, the larvae begin to swell. Sexes become distinguishable in fourth-stage larvae within the host tissue. Females continue to swell and at maturity are white, pear-shaped, and about 1–2 mm long. After the fourth stage, males become wormlike, about 1–1.5 mm long, and migrate out of the roots. Males are common in some but not all species and are functional in reproduction in some but not all species. Females remain in roots, and each may produce up to 1,000 eggs in a gelatinous matrix that is often pushed out of the root tissue. Eggs hatch in the gelatinous matrix; young larvae emerge and invade new sites of the same root or new roots. Depending upon hosts, temperature, and nematodes species, generation time is usually 20–60 days.

Epidemiology

In general, root-knot nematodes reproduce most rapidly, survive longer, and cause the most damage in coarse-textured soils. However, they are apparently limited more by temperature requirements, which vary with species, than by soil type. The "northern" root-knot nematode, *M. hapla,* has an optimum temperature of 20°C. Other species have higher temperature requirements and cannot withstand cold temperature. Hence, root-knot nematodes are of greatest economic importance in tropical and warm temperate climates and of lesser importance in northern latitudes and high elevations of tropical latitudes where soil temperatures are cool.

Because potatoes are predominately grown in the cooler climates, root knot of potato is not a major economic problem. Potato culture in warmer climates could drastically change this situation. Furthermore, environmental races exist in *M. javanica* and possibly in other species. Such strains could adapt to cooler climates and cause severe damage. For example, *M. incognita* has become well established in the midhill elevations of India and causes severe damage to potato. *Meloidogyne* spp. enhance disease development by other pathogens. On potato, development of brown rot bacteria (*Pseudomonas solanacearum*) is enhanced by *M. incognita acrita*.

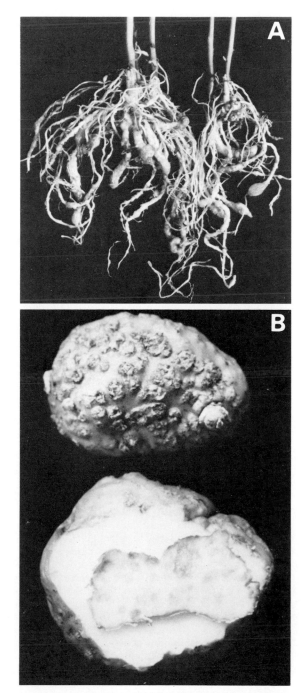

Fig. 101. Root-knot nematodes (*Meloidogyne* spp.): galls on roots **(A)** and tubers **(B)**. (Courtesy P. Jatala)

Resistance

Limited research has been done on development of potato cultivars resistant to *Meloidogyne* spp. In India, potato cultivar H-294 is resistant to *M. incognita*, and moderate to high resistance exists in several wild diploid species of *Solanum*. In Peru, resistance to *M. incognita* in hybrids of *S. demissum* was found. Also, high resistance to *Meloidogyne* spp. in *S. torvum* and partial resistance in *S. pseudolulo* and *S. quitoense* has been reported. More recently, the International Potato Center reported resistance in a number of noncultivated *Solanum* spp. In the United States, resistance to *M. hapla, M. javanica, M. incognita, M. arenaria,* and *M. incognita acrita* has been found in clones of *S. tuberosum* ssp. *andigena*.

Control

1) Because *Meloidogyne* spp. deposit their eggs in an external egg mass that is relatively unprotected, chemical control has been more successful than with cyst nematodes. In some countries where *Meloidogyne* spp. cause serious damage to potatoes, economic control has been achieved through the use of soil fumigants or the newer organic phosphate or carbamate nematicides. Dosage levels depend upon soil type, environmental conditions, and type of crop.

2) The wide host range of *Meloidogyne* spp. has made selection of suitable crops for rotation schemes difficult, although using grasses has been successful. In Rhodesia, weeping lovegrass, Katabora Rodesgrass, or Bambatsi Panicumgrass provide good control of *M. javanica*. Because of the relatively rapid decline of *Meloidogyne* spp. in the absence of a suitable host, nonhost crops can be grown for a shorter time for root-knot nematode control than that required for potato cyst nematode control.

Selected References

BRODIE, B. B., and R. L. PLAISTED. 1976. Resistance to root-knot nematodes in *Solanum tuberosum* ssp. *andigena*. J. Nematol. 8:280–281.
FRANKLIN, M. T. 1971. Taxonomy of Heteroderidae. Pages 139–162 in: B. M. Zuckerman, W. F. Mai, and R. E. Rohde, eds. Plant Parasitic Nematodes, Vol. 1. Academic Press, New York. 345 pp.
JATALA, P., and P. R. ROWE. 1977. Reaction of 62 tuber-bearing *Solanum* species to root-knot nematodes, *Meloidogyne incognita acrita*. J. Nematol. 8:290.

False Root-Knot Nematodes

Information on distibution of false root-knot nematodes is incomplete and warrants attention. This nematode is apparently native to the Andean regions of Peru and Bolivia, where losses of up to 55% are not uncommon at altitudes of 2,000–4,200 m and occasionally at lower altitudes. It also occurs in Argentia, Chile, Ecuador, the United States, Mexico, England, Holland, India, and the USSR.

Symptoms

No specific aboveground symptoms are diagnostic, although infected plants are stunted and tend to wilt under moisture stress as a result of poor root growth and/or root damage. Galls on roots are similar to those produced by root knot nematodes, and infected plants lack normal fibrous root growth. Larval invasion can cause death and deterioration of the small roots. Galls occur in a beadlike fashion, and, in Bolivia and Peru, they are commonly called "rosario," referring to rosary beads (Fig. 102).

Individual gall size depends upon nematode density, root size, and the race of the nematode. Gall shape is usually spherical and similar to that caused by *Meloidogyne arenaria*, but extension of lateral roots on the galls is usually lacking, depending upon the nematode race and host type. Although the false root-knot nematodes do not cause easily recognizable symptoms on potato tubers, they attack tubers and usually penetrate under the skin to a depth of approximately 1–2 mm.

Disease Cycle of Causal Organism

Nacobbus aberrans on potatoes has a disease cycle somewhat similar to that on sugar beets. The first molt occurs within eggs; second-stage larvae emerge, invade small roots, and establish themselves in a favorable location. Cells of the feeding site (vascular cylinder) increase in size, followed by necrosis of the cortical cells. Larvae feed and undergo two molts. They either leave the roots as preadults or continue feeding in the already established site, develop galls on roots, and complete the life cycle. A portion of those that leave the roots complete the final molt and become males or active females. Sex can be distinguished at the end of the third stage. Young female nematodes move to large roots and establish themselves with their heads near the vascular tissue. Surrounding cells enlarge and a gall develops. Posteriors of the females extend toward the outside, and an opening is formed on the root surface, where a portion of the eggs are deposited into a gelatinous matrix. Preadult and active females also invade tubers and penetrate approximately 1-2 mm below the skin surface. Although a few develop to maturity, the majority remain in a semiquiescent stage for long periods and serve as the primary means of disease dissemination. Larvae may also infect tubers, but these seldom develop beyond the larval stage and may be disseminated in that form. Depending upon host, temperature, and race of nematode, generation time is usually between 25 and 50 days.

Epidemiology

False root-knot nematodes have wide temperature adaptability, surviving and reproducing most rapidly at a temperature range of 20–26° C. Thus, they could become a limiting factor to potato production in warm climates. However, in the Andes they are associated with potatoes grown at 15–18° C and are not limited by soil type. *N. aberrans* occurring in the Andes contains two or more races differing in pathogenicity and is often found with *Globodera* and *Meloidogyne* spp. Obtaining resistance to *Globodera* or *Meloidogyne* spp. might therefore alter the competition for *Nacobbus* spp. or vice versa. *Nacobbus*-induced galls are often infected with *Spongospora subterranea*, the fungus of powdery scab.

Other Hosts

False root-knot nematodes have a relatively wide host range.

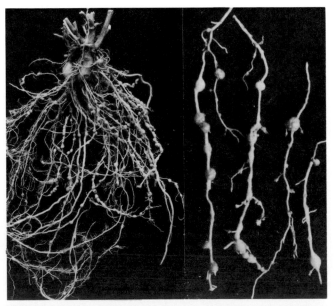

Fig. 102. False root-knot nematode (*Nacobbus aberrans*) on potato roots. (Courtesy P. Jatala)

They attack many major crops in the Andean region and many weed species, causing extensive damage to members of the *Solanaceae*. Races differ in their ability to attack certain crops.

Resistance

Limited work has been done on development of resistance to *N. aberrans*. A native *Solanum tuberosum* ssp. *andigena* cultivar shows a high level of resistance, and excellent resistance exists in *S. sparsipilum*.

Control

1) In preliminary experiments in South America, economical control has been achieved through the use of organophosphates and oxime carbamates.

2) Because of a relatively wide host range, selection of suitable crops for rotation schemes is difficult, although members of the *Gramineae* and most of the *Leguminosae* are nonhosts. Populations decline rapidly in the absence of a suitable host; therefore, rotations can be shorter than those required for the potato cyst nematode.

3) Quarantine and restriction of potato seed tuber shipments into disease-free areas should be strictly enforced.

Selected References

ALARCÓN, C., and P. JATALA. 1977. Affecto de la temperatura en la resistencia da *Solanum andigena* a *Nacobbus aberrans*. Nematropica 7:2–3.

CLARK, S. A. 1967. The development and life history of the false root-knot nematode, *Nacobbus serendipiticus*. Nematologica 13:91–101.

JATALA, P., and M. M. de SCURRAH. 1975. Mode of dissemination of *Nacobbus* spp. in certain potato-growing areas of Peru and Bolivia. J. Nematol. 7:324–325. (Abstr.).

JATALA, P., and A. M. GOLDEN. 1977. Taxonomic status of *Nacobbus* species attacking potatoes in South America. Nematropica 7:9–10.

LaROSA, D. G., and P. JATALA. 1977. Depth of penetration of *Nacobbus aberrans* in potato tubers. Nematropica 7:11.

LORDELLO, L. G. E., A. P. L. ZAMITH, and O. J. BOOCK. 1961. Two nematodes found attacking potato in Cochabamba, Bolivia. Ann. Acad. Bras. Cienc. 33:209–215.

SHER, S. A. 1970. Revision of the genus *Nacobbus* Thorne and Allen, 1944 (Nematode: Tylenchoidea). J. Nematol. 2:228–235

Lesion Nematodes

Of the several *Pratylenchus* spp. known to damage potatoes, *P. penetrans* is the most important in North America and in Europe. Species in other areas are *P. crenatus*, *P. minyus*, *P. thornei*, and *P. scribneri* in Europe; *P. crenatus*, *P. brachyurus*, and *P. scribneri* in North America; *P. andinus*, *P. scribneri*, *P. penetrans*, and *P. thornei* in South America; *P. brachyurus* and *P. scribneri* in Africa; and *P. vulnus* and *P. coffeae* in Japan.

Symptoms

High populations of lesion nematodes cause areas of poor growth; plants are less vigorous, turn yellow, and cease to grow. Damage is caused by direct feeding, and usually only cortical tissues are affected. Large numbers of nematodes cause extensive lesion formation and cortex destruction of unsuberized feeder roots (Fig. 103A). Affected roots are commonly invaded and damaged by other soil microorganisms, thus increasing root destruction. Rhizomes are not attacked as severely as roots.

In general, *P. penetrans* in primarily a root pathogen, whereas other species such as *P. brachyurus* and *P. scribneri* cause serious tuber damage. Lesions on tubers become visible when nematode numbers in a small area of the tuber are high enough to cause a number of adjacent cells to die. In South Africa, symptoms on tubers caused by *P. brachyurus* are purple-brown areas about 0.5 mm in depth, irregular in shape, and surrounded by a slightly depressed border. Raised wartlike protuberances, unsightly lesions, pimples, and weight loss and withering in storage reduce the market value of tubers and make infected seed potatoes worthless (Plate 83).

Lesion nematodes are often associated with wilt-causing fungi such as *Fusarium* spp. and *Verticillium* spp. Other fungi and bacteria are frequently present in potato tissue damaged by these nematodes.

Disease Cycle of the Causal Organism

The first molt occurs in the egg, and the second-stage larva emerges from the egg. All stages are wormlike and active. They enter plant roots usually just behind the root cap but may enter through other unsuberized surfaces of roots, rhizomes, and tubers (Fig. 103B). Entry of, and movement through, roots may be intercellular or intracellular. Entry is apparently accomplished largely by mechanical pressure and cutting action of the stylet rather than by enzymatic action. Lesion formation and root death usually occur ahead of the area penetrated. In tubers, cells surrounding nematodes are brown; cytoplasm is granular; and nuclei are reduced in size.

Males are common in some species but not in others. Bisexual reproduction occurs in species in which males are abundant. Gravid females lay eggs in the soil and roots, either singly or in small groups. Generation time is from four to eight weeks, depending upon factors such as temperature, nematode species, and host.

Epidemiology

Soil temperature requirements vary greatly with species. Optimum temperature for reproduction of *P. penetrans* is 16–20° C; it is an important pest in regions of Europe and the United States that have this temperature range. In warmer climates, species with higher temperature optima (25–28° C), such as *P. brachyurus* in Africa and *P. coffeae* in southern Japan, replace *P. penetrans*.

Damage to potatoes by lesion nematodes is usually associated with coarse-textured soils. This may be partly because some of the species involved, e.g., *P. penetrans*, are favored by sandy soils and partly because such soils are preferred for potato culture.

Soil moisture influences movement and other activities of *Pratylenchus* spp. In general, favorable soil moisture level is one

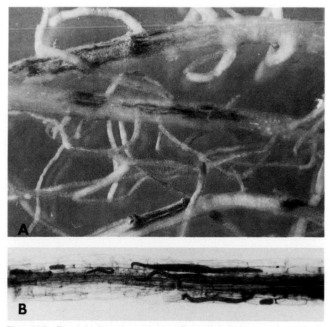

Fig. 103. Root-lesion nematode (*Pratylenchus penetrans*): **A**, damage on the roots; **B**, adults, larvae, and eggs inside root. (Courtesy W. F. Mai, B. B. Brodie, and M. B. Harrison)

at which the soil particles and aggregates are surrounded by a film of water but the intercellular spaces are free of water.

Other Hosts

Species of *Pratylenchus* that attack potatoes have wide host ranges. More than 164 hosts have been recorded for *P. penetrans*.

Resistance

A high degree of resistance in potato has not been identified, although Peconic and Hudson have some resistance. Lower population increase with certain resistant cultivars is due to fewer numbers of eggs produced per female.

Control

1) Soil fumigation decreases populations of *Pratylenchus* spp. and often increases potato yields but is practical only when yields and prices of potatoes are high. Soil fumigants are further limited because they are ineffective in fine-textured and cold soils. Nonfumigant nematicides have little or no phytotoxicity, are easier to apply, may be applied at planting time, and are more effective in a wider variety of soil types. Such nematicides are usually more practical than fumigants for controlling root-lesion nematodes.

2) Controlling root-lesion nematodes by crop rotation is not effective because the species have a wide host range. Rye is an excellent host of *P. penetrans*. Severe plant damage and yield reductions follow rye as a winter cover crop.

Selected References

DICKERSON, O. J., H. M. DARLING, and G. D. GRIFFIN. 1964. Pathogenicity and population trends of *Pratylenchus penetrans* on potato and corn. Phytopathology 54:317–322.

DUNN, R. A. 1973. Resistance in potato (*Solanum tuberosum*) to *Pratylenchus penetrans*. Abstract No. 0860:308 in: Abstracts of Papers, 2nd Intl. Cong. Plant Pathol., Minneapolis, MN.

KOEN, H. 1967. Notes on the host range, ecology and population dynamics of *Pratylenchus brachyurus*. Nematologica 13:118–124.

MILLER, P. M., and A. HAWKINS. 1969. Long term effects of preplant fumigation on potato fields. Am. Potato J. 46:387–397.

Potato Rot Nematodes

Ditylenchus destructor is primarily an important potato pathogen in the temperate regions of Europe and, especially, in the USSR, probably due to its inability to withstand drying rather than to a direct temperature relationship. It also occurs in South Africa, some areas of the Mediterranean region, a few isolated regions of North America, and in South America.

Symptoms

No specific aboveground symptoms exist. It confines its attack to underground parts of the plant, primarily the stolons and tubers but not the roots. The earliest belowground symptoms are small, white, chalky or light colored spots just below the surface of the tuber when it is peeled (Fig. 104A). Affected tissues are dry and granular. As affected areas coalesce, tissues darken and are invaded by fungi and bacteria. Tuber skin becomes paper thin and cracks as the underlying tissues dry and shrink (Fig. 104B and C). Under suitable environmental conditions in the field or in storage, bacterial wet rot may cause complete tuber destruction.

Causal Organism

The causal organism is now known as *D. destructor* Thorne (syn. *Anguillula dipsaci* Kühn, *Tylenchus devastatrix*, Kühn, *D. dipsaci* Filipjev).

Disease Cycle

Nematode inoculum may survive in the soil, on fungi, or on weed hosts, or it may be introduced by planting diseased seed tubers. *D. destructor* enters small potato tubers through lenticels or the skin near the eyes. Nematodes at first exist singly or in small numbers in the tissue just beneath the skin of the tuber, and small white lesions are present during early to midseason tuber formation. Great numbers of nematodes are present in advancing margins of the lesions where the tissue is soft and mealy. More tuber tissue becomes involved as populations increase. Tissue becomes darker in color as secondary organisms that cause dry or wet rots follow. The nematode continues to live and develop in harvested tubers. Low temperature overwintering probably occurs in the egg stage.

Epidemiology

The potato rot nematode survives in soils as low as −28° C. Greatest infestations occur at 15–20° C and at high relative humidity, 90–100%. Development occurs in the range from 5 to 34° C. The nematode cannot survive under drought or low relative humidity (below 40%). Spread of the nematode to new areas is primarily by the use of infested potato seed.

Other Hosts

D. destructor has a very wide host range of higher plants and soil-inhabiting fungi.

Control

1) Use only uncontaminated seed. All known potato cultivars are to some degree susceptible.

2) Soil fumigation is an effective control measure and can be used where it is economically feasible.

3) Crop rotation is difficult or impossible because of the wide host range of the nematode. However, rotation is reported to be an effective practice in the USSR.

4) Hot water seed treatment has not yet been developed for seed potatoes.

5) Control is difficult, but unknown factors must be operating because spread has not been extensive, and in some previously

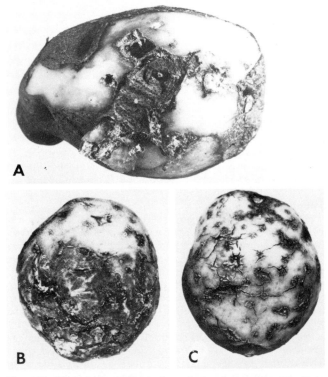

Fig. 104. Nematode (*Ditylenchus destructor*) tuber rot. Note white chalky area (**A**) and surface cracking (**B** and **C**). (A, Courtesy L. R. Faulkner and H. M. Darling; B and C, reprinted by permission of *Phytopathology*)

infested areas of North America severity of the problem has declined.

Selected References

ANDERSSON, S. 1967. Investigations on the occurrence and behavior of *Ditylenchus destructor* in Sweden. Nematologica 13:406–416.

FAULKNER, L. R., and H. M. DARLING. 1961. Pathological histology, hosts, and culture of the potato rot nematode. Phytopathology 51:778–786.

SMART, G. C., Jr., and H. M. DARLING. 1963. Pathogenic variation and nutritional requirements of *Ditylenchus destructor*. Phytopathology 53:374–381.

THORNE, G. 1961. Pages 138–148 in: G. Thorne, ed. Principles of Nematology. McGraw-Hill, Inc., New York. 553 pp.

Stubby-Root Nematodes

Stubby-root nematodes have a very wide host range in the temperate regions, and they transmit virus in many different types of plants.

Symptoms

No diagnostic aboveground symptoms exist except stunting. Roots cease elongation, resulting in numerous stunted "stubby roots" which show little or no necrosis, discoloration, or other injury symptoms.

Causal Organism

Paratrichordorus pachydermus, P. christiei, and *Trichodorus primitivus* attack potatoes.

Disease Cycle

Eggs are deposited in the soil. Immature and adult forms migrate through the soil and feed superficially on roots without becoming embedded in the plant tissues. When soil temperature is 15–20°C, the life cycle is completed in about 45 days. These nematodes most frequently occur in light, sandy soils, although they have also been reported in other soil types.

Histopathology

Feeding activity of the stubby-root nematode occurs principally at root tips. Epidermal and outermost cortical cells are punctured; as feeding proceeds, the protoplast shrinks from the cell wall. After 5–10 sec, the nematode moves on to another cell. Feeding activity is followed by a loss of meristematic activity. Parasitized roots lack a root cap and a region of elongation. Differentiation of protoxylem elements occurs almost at the root apex. Apparently, therefore, new cell production is halted but differentiation of existing cells continues.

Epidemiology

In the Netherlands, nine species of stubby-root nematodes transmit tobacco rattle virus. A close relationship exists between populations of *Trichodorus* and the virus isolates.

Tobacco rattle virus is probably not readily spread for long distances by virus-infected plant material because the strain of the virus would probably not be suitable to the nematode population of the new location. Activity of stubby-root nematodes, as determined by virus spread, is affected by soil moisture, type, and temperature. Greatest activity occurs in sandy soil at 15°C with 16.7% moisture; as soil moisture decreases, activity decreases. Very little more is known about the influence of the environment on these nematodes.

Control

1) Soil fumigants have been used to control *Trichodorus* spp. and thereby reduce spread of tobacco rattle virus.

2) Not enough is known about the host range of the nematode and the virus to advise rotation as a control measure.

Selected References

HEWITT, W. B., D. J. RASKI, and A. C. GOHEEN. 1958. Nematode vector of soil-borne fanleaf virus of grapevines. Phytopathology 48:586–595.

RASKI, D. J., and W. B. HEWITT. 1963. Plant-parasitic nematodes as vectors of plant viruses. Phytopathology 53:39–47.

ROHDE, R. A., and W. R. JENKINS. 1957. Host range of a species of *Trichodorus* and its host-parasite relationships on tomato. Phytopathology 47:295–298.

VAN HOOF, H. A. 1968. Transmission of tobacco rattle virus by *Trichodorus* species. Nematologica 14:20–24.

Nematicides

Control of nematodes in soil can be achieved through use of nematicides, of which only a limited number are presently available. All should be considered potentially hazardous, and some are difficult to apply. They must be applied correctly and under suitable environmental conditions in order to obtain their full nematicidal potential.

Dispersion through the soil and activity of most soil-applied nematicides is enhanced when soil tilth, moisture, and temperature are in the proper range. Dosage level and technique of application will depend on the nematicide, soil type, rate of control desired, and economic considerations. Nematicides may be gases, liquids, or granular solids. A well-qualified, experienced person should be consulted before applications of nematicide are attempted.

Some nematicides are fumigants, which volatilize in soil and become gases that move through soil. Others are nonfumigants, which depend upon external forces such as soil water for movement. Another category of nematicides consists of those that move systemically in the plant and can be applied to foliage.

Most currently available nematicides are either halogenated hydrocarbons, organic phosphates, or carbamate compounds. Some are phytotoxic. All are toxic to humans. Therefore, caution must be exercised in their use. Label directions must be carefully read and strictly followed.

(Prepared by W. F. Mai, B. B. Brodie, M. B. Harrison, and P. Jatala)

Aphids

Several potato virus diseases are transmitted by aphids, and identification of the vector involved is often necessary. Those that commonly colonize potato (Table III) can be easily identified by morphological characteristics (Figs. 105 and 106) that are visible to the naked eye or visible when magnified by a hand lens (Table IV). These aphids may transmit both circulative (persistent) and styletborne (nonpersistent) viruses.

Others, although seldom establishing colonies on potatoes, are vectors of some nonpersistently transmitted potato viruses (Table III). *Myzus persicae* is the most efficient aphid vector and is found worldwide.

Aphid species differ not only in morphology and ability to transmit potato viruses but also in form (morph), life cycle, and behavior, depending on the environment (temperature, relative humidity, photoperiod, and host plant condition) to which the aphid or its mother is exposed. Thus, the life cycle of any particular species may not be the same in different parts of the world. Aphids may be winged (alate) or wingless (apterous), male or female. Females may be oviparous (sexually producing fertilized overwintering eggs) or parthenogenetic and viviparous (asexually producing living young, called nymphs).

Where winter conditions are severe, most potato-infesting aphids overwinter as sexually produced eggs laid on the rough bark of a woody host or on the crown and leaves of an herbaceous biennial or perennial plant. A female nymph hatches from each egg in the spring, feeds on the expanding foliage, and after molting four times, develops into a mature, wingless, parthenogenetic female. She gives birth to female

TABLE III. Aphid Pests of Potato and Known Vectors and Nonvectors of Potato Viruses[a]

| Aphid | Potato Leafroll | Potato | | | Potato Acuba Mosaic | Alfalfa Mosaic |
		Y	A	M		
Colonizers of potatoes						
Aphis gossypii-frangulae complex[c] (melon)	−	±	−	+	+	
Aphis nasturtii[c] (buckthorn)	+	+	+	+	+	
Aulacorthum solani[c] (foxglove)	+	±	±	+	+	
Macrosiphum euphorbiae[c] (common potato)	+	±	±	+		+
Myzus ascalonicus (shallot)	+	−				
Myzus persicae[c] (green peach)	+	+	+	+	+	+
Rhopalosiphoninus latysiphon (bulb and potato)	±	−	−			
Rhopalosiphum rufiabdominalis[c] (rice root)						
Smynthurodes betae[c]						
Uncommon colonizers of or visitors on potatoes						
Acyrthosiphon pisum[c]		+				+
Acyrthosiphon primulae	−	+				
Aphis fabae[c]	+	+				
Cavariella pastinaceae[c]		+				
Hyadaphis erysimi[c]	−	+				
Macrosiphoniella sanborni		+				
Myzus certus[c]	−	+				
Myzus humuli[c]	+	−				
Myzus ornatus	+	+	−			
Nasonovia lactucae[c]	−	−			+	
Neomyzus cucumflexus	+	+	+		+	
Rhopalosiphoninus staphyleae tulipaellus	+	+				
Rhopalosiphum padi[c]		+				

[a] From Kennedy et al (1962) and Beemster and Rozendaal (1972).
[b] + = vector, − = nonvector, ± = inefficient or inconsistent vector.
[c] May overwinter as eggs.

TABLE IV. Key to Wingless Viviparous Female Aphids Colonizing Potato

I. Siphunculi absent. Cauda small, rounded. On stolons. *Smynthurodes betae.*
II. Siphunculi present. Cauda extended, not rounded.
    A. Antennae always much shorter than body; head without prominent antennal tubercles, front of head slightly convex, almost flat. Siphunculi slightly longer than cauda.
        1. Body and antennae with long hairs; green with deep orange markings between and at bases of siphunculi; siphunculi almost cylindrical, thinner just before flange. On roots. *Rhopalosiphum rufiabdominalis.*
        2. Body and antennae without long hairs; yellow to green, without deep orange markings on posterior; siphunculi cylindrical.
            a. On overwintering host, body deep green and, on potato, lemon yellow to green; siphunculi pale to light brownish with dusky tips; cauda yellow to light brown, same color as basal portions of siphunculi. On leaves and flowers. *Aphis nasturtii* (Plate 84).
            b. Body pale yellow, yellow-green to mottled blackish green, siphunculi black, cauda dusky green to black. On leaves. *Aphis gossypii-frangulae* complex.
    B. Antennae usually as long as or longer than body; head with prominent antennal tubercles; front of head broadly concave; siphunculi much longer than cauda.
        1. Siphunculi shiny black, extremely swollen; swollen part over four times as wide as the basal 1/3 and distal 1/10 part, with a well-developed flange. Body pale olive green to shiny dark olive green with a brown to black sclerotic patch covering most of the dorsum of the abdomen. On sprouts and subterranean parts. *Rhopalosiphoninus latysiphon.*
        2. Siphunculi brown to pale yellow-green; if black, then siphunculi cylindrical; on most, siphunculi only slightly swollen on distal half.
            a. Body elongated, wedge-shaped. Largest of potato-infesting aphids. Body shades of green, pink, or yellow with a darker dorsal ridge. Head with prominent, outward sloping antennal tubercles. Legs long, antennae longer than body; siphunculi cylindrical, flared outward, about 1.25 times as long as the distance between their bases. Siphunculi light brown, sometimes with darker apices. On leaves, stems, and flowers. *Macrosiphum euphorbiae* (Plate 85).
            b. Body ovoid or pear-shaped, antennal tubercles prominent and converging inward or with parallel sides.
                1.) Body pear-shaped, globular, widest just ahead of siphunculi. Head with prominent parallel, straight-sided antennal tubercles. Siphunculi straight, with prominent dark flanges at tips. Body shiny light yellow green to dark green, sometimes brownish, usually with darker pigmented areas around base of siphunculi; legs and antennae with dark joints. On leaves and flowers. *Aulacorthum solani* (Plate 86).
                2.) Body ovoid, almost same width from thorax to base of siphunculi, then sides gently rounded to abruptly meet the cauda. Head with prominent in-pointed antennal tubercles.
                    a.) Body deep pink, peach, yellowish, light green to almost colorless. Siphunculi same color as body with the tips darker; slightly swollen on apical half. Cauda short. On sprouts, foliage, and flowers. *Myzus persicae* (Plate 87).
                    b.) Body dull yellowish brown to greenish brown. Siphunculi same color as body without the tips darker, swollen towards the apex. Cauda upturned, hardly visible from above. On sprouts. *Myzus ascalonicus.*

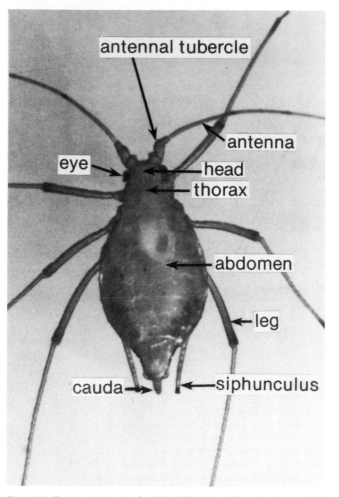

Fig. 105. Parts necessary for identification of a wingless aphid. (Courtesy M. E. MacGillivray)

Fig. 106. Outline of head region of wingless potato-infesting aphids showing shape of antennal tubercles. Left to right: *Myzus persicae, Aulacorthum solani, Macrosiphum euphorbiae, Aphis nasturtii.* (Courtesy M. E. MacGillivray)

nymphs, some of which may develop wings. Eventually, in later generations, most of the aphids will be winged and fly to herbaceous plants, weeds, and horticultural flowers or vegetables, including potatoes. Here the winged forms deposit female nymphs that, when mature, produce additional females by parthenogenesis. in late summer, when the length of the night extends past 11 hours and night temperatures decrease, sexual morphs (males and oviparous females) are produced. Winged viviparous females usually fly to the overwintering host and deposit nymphs that develop into oviparous females. In the meantime, males (winged and wingless) arrive on the host plant and mate with the oviparous females, after which the fertilized eggs are deposited to complete the cycle. Eggs are pale green at first and later become shiny black.

In mild climates, sexual aphids and eggs do not occur, and parthenogenetic, viviparous females are produced throughout the year. Some species, such as *Myzus persicae, Macrosiphum euphorbiae,* and *Aulacorthum solani,* which normally overwinter as eggs in colder climates, survive as viviparous females in sheltered places such as greenhouse and storage cellars.

Selected References

BÄNZIGER, H. 1977. Keys for the identification of aphids (Homoptera). II. Field identification of common wingless aphids of crops in Thailand. Dept. Agric. Minist. Agric. and Cooperatives, Bangkok, Thailand, and UNDP/FAO Th. A. 74/019. Plant Prot. Serv. Tech. Bull. 36:1–41.

BEEMSTER, A. B. R., and A. ROZENDAAL. 1972. Potato viruses: Properties and symptoms. Pages 115–143. in: J. A. de Bokx, ed. Viruses of Potatoes and Seed-Potato Production. Pudoc, Wageningen, The Netherlands. 233 pp.

COTTIER, W. 1953. Aphids of New Zealand. N. Z. Dep. Sci. Ind. Res. Bull. 106. 382 pp.

EASTOP, V. F. 1958. A study of the Aphidae (Homoptera) of East Africa. Her Majesty's Stationery Office, London. 126 pp.

EASTOP, V. F. 1966. A taxonomic study of Australian Aphidoidea (Homoptera). Aust. J. Zool. 14:399–592.

HILLE RIS LAMBERS, D., and M. E. MacGILLIVRAY. 1959. Scientific names of potato-infesting aphids. Can. Entomol. 91:321–328.

HOLMAN, J. 1974. Los áfidos de cuba. Inst. Cubano Del Libro, La Habana. 304 pp.

KENNEDY, J. S., M. F. Day, and V. F. EASTOP. 1962. A conspectus of aphids as vectors of plant viruses. Commonw. Inst. Entomol., London. 114 pp.

PRIOR, R. N. B., and J. R. MORRISON. 1977. Key for the field identification of brassica, potato and sugar beet aphids with photographic illustrations. Minist. Agricul. Fish. Food, London.

(Prepared by M. E. MacGillivray)

Seed Potato Certification

Because the potato plant and its tubers are vulnerable to many diseases and pests, seed quality, as determined by relative freedom from these entities, is of major importance in potato production. This became apparent during the latter part of the 19th century when pests and disease organisms were disseminated in seed stocks that were allowed to move freely throughout the world during the rapid expansion of potato culture. Because of this seriously developing dilemma, the United States established the National Plant Quarantine Act of 1912, prohibiting the importation of potatoes infected with black wart (*Synchytrium endobioticum* (Schilb.) Perc.), an embargo that is still in effect. An embargo preventing importation of potatoes from countries harboring powdery scab (*Spongospora subterranea* (Wallr.) Lagerh.) was enacted two years later.

Origin of Seed Potato Certification in North America

In 1904, while studying potato diseases in Europe at the request of the U.S. government, Professor L. R. Jones of Vermont observed seed improvement programs; he later persuaded Dr. W. A. Orton of the Bureau of Plant Industry of the U.S. Department of Agriculture (USDA) to visit Germany in 1911 to observe potato diseases and study the seed potato inspection program initiated by Dr. Otto Appel. On his return from Europe, Dr. Orton, with Dr. William Stuart of the USDA, developed a plan for the inspection and certification of potato seed stocks based upon the German system. During the next few years, they presented their proposal to growers, potato specialists, and state agricultural experiment station officials and at the annual meetings of the American Phytopathological Society and the National Potato Association of America (now the Potato Association of America).

In 1914, Dr. Orton, accompanied by Dr. Appel of Germany, Dr. H. T. Gussow of Canada, Dr. Johanna Westerdijk of Holland, and potato specialists from the USDA, visited the principal potato-growing areas of 13 states extending from Maine to California to study disease and other problems related to seed production. This study stimulated interest and added impetus to the movement for organized inspection programs. These efforts culminated in the first official Potato Seed Certification Conference, which was held in Philadelphia, PA, on December 28, 1914. Representatives from Canada, Germany, Ireland, the USDA, and 12 states participated. At this conference, the basic framework of the present seed potato certification programs was formulated.

Dr. Orton recommended that "a system of official inspection and certification" be established in each of the seed-growing states, with emphasis being placed on "freedom from disease, varietal purity and vigor." The suggestion was also made that programs be administered by a "state agency such as an Experiment Station and protected by suitable legislation penalizing misuse of certificates." The proposal, which suggested that programs be operated on a voluntary basis and that growers stand the cost of inspection, also outlined procedures for making field inspections, established disease tolerances, suggested the use of certificates and official tags for inspected seed, and proposed size and quality of tubers. Even though virus infections were not suspected at the time, the proposal recognized that "degeneration" or "running-out" of seedstocks was tuber-transmitted and suggested that "seed" tubers be selected from healthy-appearing stocks.

Seed potato certification programs in North America became a reality during 1913–1915, when Canada (New Brunswick and Prince Edward Island), Idaho, Maine, Maryland, Vermont, and Wisconsin established official programs. Ten more states started programs between 1916 and 1919, followed by three additional Canadian provinces and eight states during 1920–1922.

In the early days of certification in North America, inspectors were confronted with the problems of varietal mixtures, varietal synonyms, and degeneration or "running-out" of seed stocks. The confused picture of the degeneration complex started to clear when Quanjer et al in 1916 first established the infectious nature of potato leafroll by graft transmission; shortly thereafter, Oortwijn Botjes (1920) and Schultz and Folsom (1921) reported independently that the aphid *Myzus persicae* transmitted the causal entity of leafroll from plant to plant. Subsequent investigations exposed the involvement of numerous tuberborne virus diseases, such as mild and severe mosaic and spindle tuber (the causal agent of which is now known to be a viroid), with seed degeneration. Recognition of these diseases by certification personnel provided the basis for their subsequent control. A tremendous improvement in yield and quality of seed stocks occurred after certification personnel had mastered the diagnosis of virus diseases.

Present Day Seed Potato Certification

Seed potato certification today represents a voluntary agreement between the seed grower and the certifying agency. In the United States, programs are handled by individual states

and administered by state departments of agriculture, land grant universities, grower associations, or various combinations of these agencies. In Canada, seed potato certification is a federal program carried out by the Plant Quarantine Division of the Department of Agriculture. Since all agencies have been given official status by their respective state or national governments, protection against the fraudulent use of the term "certified seed potatoes" has been assured. Disclaimer clauses limiting liability to the value of the seed are commonly used by all agencies to protect both the grower and the agency.

Each agency publishes certification standards outlining eligibility requirements for inspection, grower fee schedule, disease tolerances, grade requirements, winter tests, and rules governing sale of certified seed. Application inspection forms are sent to individual growers before the planting season requesting a listing of cultivars, classification and source of seed lots, acreage, field numbers and location, previous crop history of fields, and date of planting. This information is used by the agency to determine eligibility for inspection.

Fields are planted with pretested and approved seedstocks. Practices are adopted that minimize the spread of viruses and of pathogens borne by soil and debris. Procedures for inspections vary among states but, in general, are quite similar. A minimum of two field inspections is made during the growing season at a time most opportune for detecting diseases and varietal mixtures by visual examination. Field tolerances for disease and varietal mixtures vary among agencies but, in general, are quite similar. Those currently used in Wisconsin for the Certified class are fairly typical (Table V).

Failure to meet the tolerances is cause for rejection and, in addition, certification may also be denied for the prevalence of other diseases such as blackleg, haywire, or wilts. Lack of isolation, unsuitable cultural conditions, high aphid populations, nematodes, unsatisfactory performance of test samples, or other factors that may impair seed value may also be cause for rejection. Of the acreage rejected from 1968–1972 in North American agencies certifying over 1,000 acres per year, bacterial ring rot, leafroll, varietal mixture, mosaic, and blackleg accounted for 5.6, 0.8, 0.7, 0.5, and 0.4%, respectively.

Harvest inspections are required by some agencies, whereas others require both a harvest and a bin inspection, which identifies the stored seed and includes estimates of volume, grade, and tag eligibility. Certification is not complete until the seed has been graded for quality and size in conformity with seed grades and identified with official tags and seals and has passed inspection by the State-Federal Inspection Service. Inspection reports are issued to the grower by the inspector after each field, harvest, and bin inspection.

Most agencies in North America certify two basic classes of seed potatoes—"Foundation" and "Certified." Requirements for the production of Foundation class seed are much more rigid, e.g., Foundation class seed growers must enter their entire acreage for inspection; field tolerances for disease are approximately one fourth of those allowed for Certified classes; and requirements for seed source, land, isolation, sanitation, and the handling and storage of the crop are much more stringent. This class of seed is used to plant certified seed potato fields. Many certifying agencies require a winter test as part of their Foundation class requirement. A few agencies have made it compulsory for all seed lots, irrespective of class. These tests, which are conducted in Alabama, California, or Florida, where growing conditions favor symptom expression of diseases, consist of field planting 300–800 tubers per lot and reading the resulting plants for disease content, primarily virus, and other factors pertinent to seed productivity. To receive Foundation designation, most agencies require that a seed lot shall not show a total in excess of 0.5% of the diseases mosaic, leafroll, and spindle tuber in the winter field test.

Because of the short interval between harvest and planting of the winter test in Florida, dormancy of certain cultivars must be broken. Rindite, a 7:3:1 mixture of ethylene chlorohydrin (2-chloroethanol), ethylene dichloride (1,2-dichloroethane), and

TABLE V. Field Tolerances (%) for Certified Potatoes, as Required by Wisconsin Certified Seed Potato Agency, Madison

Disease or Varietal Mixture	Inspection	
	First	Second or Subsequent
Leafroll	1.5	1.0
Mosaics	2.0	1.0
Spindle tuber	1.0	1.0
Total	3.0	3.0
Bacterial ring rot	0.0	0.0
Varietal mixture	1.0	0.1

carbon tetrachloride, applied at the rate of 141 ml/m³ (4 cc/ft³) of treatment chamber (container), is one of the most effective chemicals used. The total dosage is applied at 24-hr intervals over a three-day period in an air-tight chamber filled to not more than one-third to one-half of its total volume. Samples are stacked at a uniform height of approximately 4 ft (1.2 m) on 6-in. (15-cm) pallets arranged in rows that allow for ample air circulation provided by fans. Before treatment, samples are warmed for five days at 24–25° C, and during treatment temperature is maintained at 25–26° C; higher temperatures may result in injury. Rindite is placed in pans above the potatoes, and burlap bags are arranged to dip into the chemical and act as wicks. Rindite is highly toxic to humans and proper safety precautions must be taken.

Certified seed is packed and shipped in clean, new bags or in bulk in clean, disinfested carriers. Individual bags and carriers containing bulk shipments are tagged, indicating cultivar, crop year, serial or certification number, and grower's name and address. Tags are attached to each container so that neither can be opened without breaking the seal.

In the United States all certified seed potatoes are graded in conformity with standards established by individual states; in Canada they are established by the federal government. Grades vary among states but basically are quite similar to the federal grade of "U.S. No. 1 Seed Potatoes" established in 1972, which serves as a reference point for marketing seed potatoes, replacing the U.S. No. 1 table grade. Most certifying agencies in North America have several grades, with a blue tag most frequently representing top quality and other colors for grades with less stringent standards. Size restrictions, which can vary if specified, usually range from 1½–3¼ in. (3.8-8.3 cm) in diameter. with a maximum weight of 12 oz (340 g).

In the United States, fees for certification services are borne by individual growers, whereas in Canada they are absorbed by the federal government. Fees cover the cost of application, field inspections based on acreage, virus tests, shipping point inspections for individual bags or bulk shipments, winter tests, tags and seals.

Following completion of field and harvest inspections, agencies publish crop directories listing growers and all varietal acreages that have met certification standards. Directories are also released by agencies following completion of the southern winter tests. Both of these publications, which are widely distributed within the potato industry, serve as an important tool in locating reliable seed sources.

Seed potatoes in North America are produced primarily in the northern states along the Canadian border, in areas of high elevation in certain western states, and in all Canadian provinces except Newfoundland. At one time as many as 36 states certified seed potatoes. However, the rapid spread of virus diseases was the primary reason that southern states dropped out of seed production. The spread of virus is considerably less apt to occur in northern areas, primarily because of lower populations of viruliferous insects, such as the green peach aphid, which spread many of the viruses that attack potato.

From 1968 to 1977, approximately 78% of the total seed production in the United States was produced in Maine (25%), Idaho (22%), North Dakota (17%), and Minnesota (14%). In 1977, 19 states produced a total of 230,458 acres of certified seed potatoes, of which these four states accounted for approximately 77%. Prince Edward Island and New Brunswick are the leading seed-producing provinces in Canada, accounting for approximately 86% of the total acres (66,888) passing certification in 1977. During the past 10-year period, they have produced approximately 84% of Canada's seed acreage.

Seed Improvement Programs

Seed improvement programs are constantly striving to upgrade the quality of nuclear (elite) seed stocks, which serve as the basis for certified seed potato production.

In an attempt to more effectively control several virus diseases and bacterial ring rot, four states and four Canadian provinces have established official Foundation (elite) seed farms where nuclear seed stocks are developed for their respective seed industries. The practice of planting cut seed in North America favors rapid spread of several diseases because the causal organisms are easily transmitted mechanically; such spread is curtailed by the planting of whole seed.

These official seed farms are located in well-isolated areas that have a history of low insect populations and are staffed with personnel having the expertise to perform the technical procedures required for development of nuclear seed stocks. Practices involving strict sanitation, application of systemic insecticides at planting, and rigid spray schedules for control of insects and foliar diseases are adhered to at all times.

In recent years, considerable time and effort have been devoted on these farms to developing nuclear seed stocks free from such latent viruses as X, S, and M. Virus-free programs have been developed primarily because of potato virus X (PVX), which causes latent mosaic. PVX, known as the "healthy potato virus" because of its symptomless characterisitics, produces visible symptoms only under certain environmental conditions, making control by roguing extremely difficult. Partly for this reason, seed potato certification personnel welcomed the introduction of virus X-free certification programs; they fully appreciated that the developing virus X-free projects of the 1930s in Europe, which led to such official programs in America after 1945, were big steps forward in seed potato improvement. In North America, seed specialists are not in complete agreement as to the desirability of maintaining totally virus-free seed stocks; some feel that advantages exist in the cross-protection provided by mild strains of PVX.

Procedures for the development of nuclear seed stocks vary on these official seed farms, but in general, programs are based on clonal selections. A clone is a stock of tubers or plants derived from the same mother plant by vegetative propagation. Clonal selection implies the increase of stocks from selected healthy plants of desirable varietal type and their subsequent multiplication.

In recent years, meristem and shoot tip culture have been used to obtain virus-free stocks of standard cultivars, old cultivars of historical interest, and promising new seedlings. This procedure is based on the fact that cells in the growing tips of axillary buds, as well as in tips of sprouts, may, by exposure of the plant or tuber to high temperatures, "grow away" from viruses even though the plant or tuber is systemically infected. By removing the meristem tips and allowing them to develop on special media, virus-free plantlets can be obtained. In practice, a high percentage of success in obtaining virus-free plants has been achieved when meristem tips have been taken from rooted stem cuttings or sprouting tubers that have been exposed to 35–38° C for 4–6 weeks before bud excision.

Plantlets produced in vitro by meristem tissue culture are not necessarily pathogen-free and so should be thoroughly screened for freedom from bacteria, fungi, viruses, and viroids. Some of the diseases or pathogens for which screening is conducted and the testing procedures used to detect them are: bacterial ring rot (eggplant and tomato as indicator plants and the broth test for plantlets produced in vitro); potato virus X (*Gomphrena globosa* as indicator plant and serology); potato virus S (serology); potato virus A (plant indicator 'A6') and potato virus Y (serology, indicator plants 'A6,' and *Solanum demissum* P.I. 230579); leafroll (aphid transmission to indicator plant *Physalis floridana*); and spindle tuber (polyacrylamide gel electrophoresis). In the latter test, the pathogen (viroid) is a ribonucleic acid that differs significantly from the nucleic acids that occur in healthy plants. Separation by electrophoresis and subsequent staining permits reliable detection of the viroid, including strains that cause no visible symptoms.

Stem cutting, which was developed as a means of eliminating bacterial and fungal pathogens normally carried over by tuber propagation, is a relatively new procedure that has proven to be a valuable tool in seed improvement. Plants from selected tubers are grown in the greenhouse and then topped when they are

15–40 cm high so as to stimulate the growth of axillary shoots. These shoots, which are used as cuttings, are removed when they are 5–7 cm long (treating with a rooting hormone is optional) and transplanted to moist, sterile sand or vermiculite (held under intermittent mist in some institutions) until rooting occurs, usually 10–12 days. Following an optional hardening-off period in a cold frame, they are transplanted to the field. In the meantime the "mother" plants, from which the cuttings were taken, are screened thoroughly to assure freedom from the various pathogens. In 1970, Scotland became the first country to require that all seed lots entered for certification be derived from nuclear stocks developed from virus-tested stem cuttings.

Once disease-free material has been obtained, procedures for further increase of nuclear seed stocks on official seed farms vary; however, a thorough indexing program (greenhouse and winter field test), utilizing various combinations of tuber, hill, and tuber-unit indexing, is used in all programs. These stocks are increased primarily as four-cut tuber units (four consecutive hills per tuber) for two years before being sold to foundation growers. During this increase period, they are rogued intensively during the growing season and screened thoroughly for the presence of viruses. Foundation seed growers are encouraged to use a "flush-out" system that entails purchasing nuclear stocks each year from official seed farms and increasing them for no more than 2–3 years.

Certifying agencies, which do not have official seed farms, must rely on a few selected growers for the increase of nuclear seed stocks.

The production of top quality seed potatoes has become a highly sophisticated and technical procedure that undoubtedly will become more complex as new technology is developed. It is imperative that all those associated with seed production be receptive to technological improvements to ensure continuous progress in the seed industry.

Selected References

DARLING, H. M. 1977. Seed potato certification. Pages 405–416 in: O. Smith, ed. Potatoes: Production, Storing, Processing, 2nd ed. Avi Publishing Co., Westport, CT. 776 pp.

HARDIE, J. L. 1970. Potato Growers' Guide to Clonal Selection. McCorquodale Ltd., Glasgow, Scotland. 20 pp.

MELLOR, F. C., and R. STACE-SMITH. 1977. Virus-free potatoes by tissue culture. Pages 616–646 in: J. Reinert and Y. P. S. Bajaj, eds. Applied and Fundamental Aspects of Plant Cell, Tissue, and Organ Culture. Springer-Verlag, New York. 803 pp.

MOREL, G., and C. MARTIN. 1955. Guérison de pommes de terre atteintes de maladies á virus. C. R. Hebd. Séanc. Acad. Agric. Fr. 41:472–474.

MUNRO, J. 1954. Maintenance of virus X-free potatoes. Am. Potato J. 31:73–82.

OORTWIJN BOTJES, J. G. 1920. De bladrolziekte van de aardappelplant. M.S. thesis, Landbouwhogeschool, Wageningen. 136 pp.

ORTON, W. A. 1914. Inspection and certification of potato seed stock. Phytopathology 4:39–40 (Abstr.).

QUANJER, H. M., H. A. A. VAN DER LEK, and J. G. OORTWIJN BOTJES. 1916. Aard, verspreidingswijze en bestrijding van phloemnecrose (bladrol) en verwante ziekten. Sereh. Meded. Landbouwhogesch. 10:1-138.

RIEMAN, G. H. 1956. Early history of potato seed certification in North America, 1913–1922. Potato Handbook. Vol. 1, pp. 6–10. Potato Assoc. Am., New Brunswick, NJ.

SCHULTZ, E. S., and D. FOLSOM. 1921. Leafroll, net-necrosis, and spindling-sprout of the Irish potato. J. Agric. Res. 21:47–80.

SEED REPORT. 1977. Spudlight. Certified Seed Edition, Part 2. United Fresh Fruit Veg. Assoc., Washington, DC. 15 pp.

(Prepared by E. D. Jones, James Munro, and H. M. Darling)

Key to Disease

Diagnostic Microbial Structures

Sclerotia
Black—white mold, gray mold, Rhizoctonia, charcoal rot
Purplish black—violet root rot
Very small, black—black dot
Tan stem rot

Bacterial Exudate from Vascular Tissue
Abundant, gray—brown rot
Sparse, white—ring rot

Prominent Mycelium in or on Soil
White—white mold
White, fanlike—stem rot
Grayish white—Rosellinia
Brown, in strands near or on tubers—Xylaria, Armillaria

Entire Plant

Distinct Symptoms Lacking
Unusually large—giant hill
Off-color, poor growth—general nutrient deficiency, several nutrient imbalances
Bronzing
   General or of old leaves—photochemical oxidant air pollution, potassium or zinc deficiency, Fusarium wilt
   Of tip leaves—psyllid yellows
Upright growth habit—PLRV in andigena types, PSTV

Stunting and/or Poor Growth
Beginning in localized areas of field nematodes (cyst, root-knot, false root-knot, stubby-root, lesion).
Mild to severe—several virus diseases (PLRV, PVY, PVM, APMV, TBRV, PSTV, PYDV, BCTV)
   Pale light green to yellow leaves, later necrotic—magnesium deficiency
   Dark green—several mosaic viruses
   Early in season—potassium deficiency
   Lusterless—phosphorus deficiency
   Dwarfing—BCTV, PYDV, PMTV
Early maturity—several nutritional deficiencies
   Lower leaves mature, speckled—photochemical oxidant air pollution
   General chlorosis, usually from base upward—Verticillium wilt
Late maturity, larger than normal—genetic abnormalities

Wilt
With drought stress—nematodes (cyst, root-knot, false root-knot)
Chlorosis, wilt, and early death—black dot, violet root rot, brown rot, charcoal rot
   In groups in the field—Rosellinia
   With vascular discoloration—Verticillium wilt, Fusarium wilts
Wilt or chlorosis
   Of one stem or on one side of leaf, stem, or side of plant—brown rot, Fusarium wilt, Verticillium wilt, ring rot
   Of green leaves and stems, later chlorosis and necrosis—ring rot, brown rot, pink rot, stem rot
   Rapid green collapse and death of some or all plants in localized areas of field—lightning injury

Necrosis
Systemic (top necrosis)—PVX, PAMV, APMV, PVT, TSWV
Defoliation from bottom toward top of plant—early blight, photochemical oxidant air pollution, powdery mildew, pink rot, PVY

Deformed
Internodes shortened
   Downward leaf curl—potassium deficiency
   Bushy appearance—boron deficiency, PYDV, Fusarium wilts
Spindly, crinkled leaves—chemical injury
   Chlorotic to necrotic margins—phosphorus deficiency
Numerous stems, bushy—genetic abnormalities, mycoplasmas
Rosette
   At tip—calcium deficiency
   With dwarfing—ring rot
Red pigmentation or chlorosis of apical leaves, stunting, general chlorosis, thickening of nodes, aerial tubers, vascular necrosis, basal stem necrosis, or early death—blackleg, Rhizoctonia, mycoplasma, Fusarium wilts, BCTV, psyllid yellows

Stem

Numerous—coiled sprout
And thin—genetic abnormalities, hair sprout, mycoplasmas, TMV

Stunting
Of some stems—TRV
   And internodes shortened—PVY, PMTV, PYDV

Lesions Originating Below Ground
Black, extending upward from seed tuber, pith dark, cortical decay—blackleg, pink rot
Brown—common scab, skin spot, black dot
   Girdling of stem—Rhizoctonia
Rot, with dry shredding—Fusarium wilts, mycoplasmas
With prominent gray fungus, mold growth—Rosellinia

Lesions at or Near Soil Surface
Girdling and collapse, bleached color—heat injury
White epidermis at soil line, cross-hatched collapse of pith—lightning injury
No underlying necrosis on stem surface
   White mycelium—*Rhizoctonia solani* (perfect stage)
   Red to purple mycelium—violet root rot
Surface moldy, white at soil line, watery rot, stem collapse
   Black sclerotia—white mold
   Tan sclerotia—stem rot

Lesions Above Soil Surface
Epidermis white, underlying tissue usually unaffected—sunscald, hail injury
Necrosis, black to brown
   Extensive—late blight
   Restricted in size—early blight
   Stippled flecks—powdery mildew
      Also of petiole, stem brittle—manganese toxicity
Brown necrotic streaks on stems and petioles—PVY, TSWV
   At petiole attachment—blackleg, late blight, gray mold
   Light brown broad zonation—white mold

Vascular Discoloration
Brown, most severe in lower part, extending also above ground—Fusarium wilts, Verticillium wilt
Black—brown rot, blackleg

Pith Necrosis
Brown, near tip—PYDV
Black at base—blackleg
At nodes—Fusarium wilts

Aerial Tubers and/or Enlarged Nodes—pink rot, Fusarium wilts, mycoplasmas, psyllid yellows, Rhizoctonia

Galls—wart, smut

Swelling, Twisting, and Deformation—rusts, coiled sprout

Leaves and Leaflets

Mosaic Mottles
Symptomless, very mild to severe rugosity—PVX, PVS, PVM, PVY, PVA, PVT, APMV, APLV, CMV, TMV, TRV, AMV, PAMV, deforming mosaic, frost injury
On tip leaves—Fusarium wilts
Netting of minor leaf veins—APLV
    And veinal necrosis—PVT
Rough, crinkly—PVY, PVM, PYVV
Yellow areas, pale to bright—AMV, TRV, PMTV, TBRV, PYVV, TRSV
    More severe on lower leaves—PAMV

Chlorosis
General—Fusarium wilt, Verticillium wilt
Greenish spots on low shaded leaves—PVS

Interveinal Chlorosis or Necrosis
Of tip leaves—manganese deficiency
Of lower leaves—ring rot
To white yellow chlorosis—sulfur oxide air pollution, manganese toxicity, magnesium deficiency

Deformed
Crinkled, rugose—PVY, PVA, deforming mosaic
Small, twisted, possibly with short petioles and stem internodes—PAMV, PVM, TMV, CMV, TBRV, TRV, PMTV, APMV, PSTV, genetic abnormalities, chemical injury
Young leaves twisted and cupped—zinc deficiency
Cupped—phosphorus deficiency
    On tip leaves—TBRV
Irregular holes, banded, mottled—low temperature vine injury
Elongate, puckered, veins prominent—chemical injury, CMV, TMV, low temperature nitrogen toxicity
Pustules, orange, red to brown enlargements, or twisting—rusts
Small with fluted margins, acute petiole-stem angle—PSTV
Small, numerous thin stems—mycoplasmas
    Many leaves, simple (not compound)—genetic abnormalities, witches' broom, stolbur

Upward Rolling
Stiff, papery texture, pale color—PLRV, boron deficiency
Throughout plant, downward roll of petioles—PYDV
Of tip leaves, possibly pink at margins—PLRV, manganese deficiency, PVM, mycoplasmas, BCTV, psyllid yellows
With chlorosis—zinc deficiency
Of lower leaves or throughout the plant, chlorosis absent to mild—nonvirus leafroll, PLRV, PVM, potassium, phosphorus, or boron deficiency
    Chlorotic or necrotic margins—calcium deficiency
Upward rolling severe at plant tip, chlorosis or red at bases of tip leaves, may be accompanied by aerial tubers—blackleg, Rhizoctonia, mycoplasmas, Fusarium wilts
Dwarfing, marginal and interveinal chlorosis—PLRV in andigena types
Thick, brittle, interveinal chlorosis or necrosis—magnesium deficiency
Tattered edges or with holes—wind or hail injury, Ulocladium blight

Necrosis
At tip of plant—frost injury

Bronzed
    On upper surface and margins—photochemical oxidant air pollution
    Necrotic—potassium deficiency
    And necrotic spots—PVS
    And extensive systemic necrosis—Fusarium wilt
On epidermal surface or extending through the leaf—wind injury
Of veins—PVT
    And stem streak—PVY
    Flecks to streaks—manganese toxicity
Systemic leaflet and petiole necrosis—AMV, APMV, TSWV
    And leaf drop—PVY

Necrotic Lesions
Speckling of lower leaves but also on upper leaves—nitrogen deficiency chemical injury, photochemical oxidant air pollution, manganese toxicity
Necrotic spots and rings—TBRV, TRSV, TSWV, PVY
Necrosis of petioles and leaflets—TSWV, PVY
Without concentric zonation
    Black when wet, brown when dry, with or without sparse white sporulation, possibly yellow halo—late blight
    Initially water-soaked large necrotic lesions—Choanephora blight
    Yellowish to purple—Cercospora leaf blotch
    Tan, angular—*Stemphylium consortiale*
    At mechanical wounds, dark to black—Ulocladium blight
    Green to black, white to gray-brown sporulation—powdery mildew
    Light colored—*Pleospora herbarum*
With concentric zonation
    Broad zonation, wedge-shaped, circular to irregular, tan sporulation—gray mold
    Narrow zonation
    Round, brown—Phoma leaf spot
        Round or angular lesions with pycnidia—Septoria leaf spot
        Brown black—TSWV
            Angular (limited by veins)—early blight, *Alternaria alternata*

Tuber

Small Size—various viruses, genetic abnormalities, nutrient imbalances
Many tubers—witches' broom, psyllid yellows, second growth

Flaccid or Wilted—mycoplasma

Forming Secondary Tubers or Plants Prematurely—second growth, secondary tubers

Galls
Green, brown to black—wart
Tuberlike, deformed—smut
Raised, pimplelike, purple-brown—powdery scab, skin spot
White tufts—enlarged lenticels
    To brown galls later becoming necrotic depression—powdery scab
Warty—root knot nematode
    Also pimples—lesion nematode

Deformed
Irregular shape—second growth, compacted soil, PAMV, PSTV, Rhizoctonia, mycoplasmas
Set close to stem, stolons very short—Rhizoctonia, blackleg, mycoplasmas
Protruding eyes—second growth, PYVV
Pointed ends—second growth, PSTV
Elongate, round in cross-section—PSTV
Dwarfed, deformed, possibly cracked, with internal necrotic spots—AMV, PYDV, TRV, TSWV, PSTV, chemical injury
Internal arcs or rings—TRV, PMTV
Warty with internal black, scattered areas—smut

Normal Tuber Shape, Surface Unblemished
Black sclerotia on tuber surface (soil that will not wash off)—Rhizoctonia
Interior glassy or watery throughout or at stolen end—second growth, frozen tissue
Starch deposition irregular to very low—second growth, immature tubers
Sugar in tissue—low temperature storage

Surface and/or interior
  Shades of green—tuber greening
  Underlying tissue collapsed—sunscald
Subsurface tissue
  Blue to black—blackspot
  Dried, soft or sunken—wind injury
  Brown at stolen attachment—stem-end browning
  Necrotic arcs to flecks—TRV, PMTV
Center, hollow or cracked—hollow heart
Interior firm
  Necrotic (rust colored) spots or flecks principally in medullary
    tissue—internal heat necrosis, phosphorus deficiency, calcium
    deficiency, AMV, PAMV, PYDV, TRV
  Necrotic to tan discolored areas—PVY^c
  Rust colored arcs or rings—TRV, PMTV
  Net (phloem) scattered necrosis—PLRV, low temperature injury,
    stem-end browning, PAMV
  At stolon attachment—stem-end browning, chemical injury, low
    temperature injury, calcium deficiency, Fusarium wilt, Verticillium
    wilt
  Mahogany colored interior—low temperature injury
  With glassy texture, starch depletion—second growth
Interior firm to soft
  Smoky gray discoloration—low temperature injury, leak
  Black medulla—blackheart
Vascular discoloration
  Soft texture, ooze when squeezed—ring rot, brown rot
  Firm to soft, dark—low temperature injury
  Firm, confined to vascular area, brown to black possibly with water-
    soaked border—Fusarium wilts
  Firm near stolon attachment—calcium deficiency, stem-end
    browning, Fusarium wilts
  Netlike, more severe near stolon end—Verticillium wilt, chemical
    injury, stem-end browning
Discoloration evident through skin
  Indistinct lines, arcs, or blotchy areas—TRV, PMTV, TSWV,
    PAMV
    As raised rings—PMTV
  Gray-brown discoloration—brown rot, pink rot, pink eye

Normal Tuber Shape, Surface Blemished Without Active Rot
Chalky white spots below skin, later dry granular, becoming
  darker—rot nematode
Blisters—TNV
Brown rings to necrotic areas—PVY^c, PMTV, TRV
Skin cracked—tuber cracks, Rhizoctonia, PYDV, PSTV, PMTV,
  TNV, boron deficiency
  Or feathered—surface abrasions
Lenticels affected—enlarged lenticels, bacterial soft rot, powdery scab,
  stem rot, charcoal rot, gangrene, PSTV
  Underlying tissue brown to black—pink rot
Stolon attachment discolored—chemical injury, jelly end rot, stem-end
  browning, brown rot, ring rot, stem rot, Verticillium wilt,
  Fusarium wilts, boron deficiency, charcoal rot, pink rot
  Also of medullary areas—soft rot, blackleg
Eyes discolored
  Principally at tuber tip—pink eye, Verticillium wilt
  Anywhere—charcoal rot, Fusarium wilts
  Dark colored—gangrene
    With soil adhering—brown rot
  Purplish black raised spots—skin spot
  With watery exudate—frozen tubers
Wounds infected—bacterial soft rot, powdery scab, leak, Fusarium
  tuber rots, gangrene
Shallow lesions anywhere
  Necrotic skin, dark—gangrene, sunscald, high temperature field
    injury, low temperature injury, wind injury, TNV
  Silvery sheen to light brownish surface—silver scurf
  Silvery to brown with very small black sclerotia—black dot
  Irregular purplish brown—lesion nematodes
Raised lesions
  Purplish corky to necrotic spots, sometimes around eyes—skin
    spot
  Circular lesions—Thecaphora smut
Reddish purple mycelium and sclerotia on surface, sunken areas
  below—violet root rot
Brown sunken blotches—PAMV
Brown to dark sunken pits, russet or raised lesions or russet
  discoloration
  Underlying tissue firm, corky, at stolen end—potassium deficiency

Tan brown to black—common scab, TNV
Raised pustules white, later depressed, dark brown—powdery scab
Pitted—Rhizoctonia
Necrotic often in bands, scablike—chemical injury
Firm lesions, slightly sunken, relatively shallow
  Reddish brown—late blight
  Black—early blight
  Purplish mycelial mat—violet root rot
  Circular—Fusarium wilts, Fusarium tuber rot

Active Rot
Skin cracks, discolored—ring rot
Cavities spongy, shrunken—Fusarium tuber rots
  With white mycelium and black sclerotia—white mold
Cortex less severely rotted than medulla—ring rot leak
  Possible holes through the tuber—lightning injury
Gray smoky discoloration
  Of cortex, vascular ring, or medulla—low temperature injury
    Black medulla—blackheart
  Anywhere bordering rotted tissue—leak
Watery
  At stolon end—jelly end rot
  On exposure to air shades of pink, brown to black—frost injury
    Charcoal rot, later spongy—leak, pink rot
  Later chocolate brown dries as zonate lesions—Rhizopus
Semiwatery
  Brown flabby decay—gray mold
  Light gray, cavities with mycelium, and black sclerotia—charcoal rot
Advancing margin clearly delimited—white mold, Fusarium tuber rot
  Dark line—leak, pink rot
Tissue firm to dry, punky with cavities
  Pustules of spores on surface—Fusarium tuber rot
  Dry granular, shrunken—rot nematode
Vascular tissue discolored reddish brown to black—brown rot, ring rot
Lesions thumbnail to deep, dark cavities—gangrene
Loose gray white mycelium on surface and in soil, rot black,
  carbonaceous—Rosellinia
Fanlike mycelium on tuber and on soil surface, semifirm decay, cheesy
  rot—stem rot
Tuber shell remains—leak

Secondary Rots
Slimy, cream to tan, often foul odor—bacterial soft rot, Fusarium
  tuber rots. These and others follow primary pathogens such as brown
  rot, ring rot, low or high temperature injury, blackheart, late blight,
  stem rot, charcoal rot, Rhizopus rot, and rot nematode.

Seed Tubers

Decay—oxygen relations, blackheart, adverse temperature (low or
  high), Fusarium tuber rots, leak, bacterial soft rot, stem rot
Dormant, or delayed in germinating—PYDV, APMV, mycoplasmas
Dormancy lacking—psyllid yellows, witches' broom, second growth
Abnormally thin—genetic abnormalities, hair sprout, mycoplasma
  Many—witches' broom
Growing into the tuber—internal sprouting
Forming tuber directly—second growth, secondary tubers, hair
  sprouts, calcium deficiency
Swollen, curved—coiled sprout
Necrotic
  Just below tip—calcium deficiency
  Soft, progressing from seed piece—blackleg
Brown lesions
  In storage, transverse cracks—skin spot with girdling
  In field—Rhizoctonia

Stolons

Short—potassium deficiency
Numerous, long—genetic abnormalities
Dark, covered by gray-white mycelium—Rosellinia
Necrosis—pink rot
Brown lesions—skin spot
  And girdling—Rhizoctonia
  With dotlike sclerotia—black dot

Galls
    White to brown—powdery scab, smut
    On tips—wart

Roots

Galls
    White to dark brown—powdery scab, root-knot nematode
    As beads along root—false root-knot nematode
Cysts, white to brown—cyst nematodes
Brown lesions—skin spot
Cortical decay

Dotlike sclerotia—black dot
    Large sclerotia—Rhizoctonia
Cortical injury—lesion nematodes
General rot—Fusarium wilts
Necrosis
    Brown to black—pink rot
    Dark, covered by gray white mycelium—Rosellinia
Poor development—potassium deficiency, phosphorus deficiency, nitrogen toxicity
Stunted—magnesium deficiency, boron deficiency, aluminum toxicity, cyst nematodes, stubby-root nematode
Proliferation of lateral roots—cyst nematodes, root-knot nematode

(Prepared by W. J. Hooker)

Equivalent Names of Potato Diseases

Common Name / Causal Factor	Other Names	Spanish	German	French
Air pollution injury / Photochemical oxidants / Sulfur oxides		Daños provocados por contaminación ambiental / Oxidantes fotoquímicos / Gases sulfurosos		
Alfalfa mosaic / virus / AMV	Lucerne mosaic / Calico / Tuber necrosis virus	Mosaico de la alfalfa	Kalikokrankheit	
Alternaria alternata				
Andean potato / latent virus / APLV		Virus latente de los Andes		
Andean potato / mottle virus / APMV		Moteado andino		
Armillaria dry rot / *Armillaria mellea*			Hallimasch	Pourridie-agaric / Armillaire
Aster yellows, stolbur, and allied diseases / Mycoplasma	Purple top wilt / Tomato big bud / Purple top roll / Haywire / Late breaking virus / Blue stem / Moron / Purple dwarf / Yellow top / Bunch top / Apical leafroll	Punta morada / Amarillamiento apical violáceo	Stolburkrankheit / Parastolbur / Metastolbur	
Bacterial soft rot / *Erwinia carotovora*	Soft rot	Pudrición blanda	Knollennassfäule / Nassfäule	Pourriture molle bactérienne
Black dot / *Colletotrichum atramentarium*		Antracnosis / Punteado negro	Colletotrichum-Schalennekrose / Colletotrichum-Welke	Dartrose / Anthracnose
Blackheart		Corazón negro	Schwarzherzigkeit	Coeur noir
Blackleg		Pierna negra	Schwarzbeinigkeit	Jambe noire
Blackspot	Internal bruising / Enzymatic graying / Blue bruise / Bluespot	Mancha negra no infecciosa	Graufleckigkeit / Schwarzfleckigkeit / Blauverfärbung	Taches cendrées / Tachetures bleues / Taches plombées
Brown rot / *Pseudomonas solanacearum*	Bacterial wilt / Southern bacterial wilt	Marchitez bacteriana / Pudrición parda	Schleimkrankheit / Bakterielle Braunfäule	Pourriture brune
Cercospora leaf blotches / *Cercospora* spp.	Cercospora leaf spots	Manchón foliar	Gelbfleckigkeit / Cercospora-Blattflecken-krankheit	Taches foliaires / Cercosporiose
Charcoal rot / *Macrophomina phaseoli*		Pudrición carbonosa		
Chemical injury		Daños por agentes químicos		
Choanephora blight / *Choanephora cucurbitarum*		Muerte regresiva		
Coiled sprout		Brote doblado		

(continued on next page)

Common Name Causal Factor	Other Names	Spanish	German	French
Common rust *Puccinia pittieriana*		Roya común		
Common scab *Streptomyces scabies*	Scab	Sarna Sarna común	Kartoffelschorf	Gale commune
Cucumber mosaic CMV		Mosaico del pepinillo	Gurkenmosaikvirus	
Deforming mosaic PDMV		Mosaico deformante		
Deforming rust *Aecidium cantensis*		Roya peruana		

Early blight
Alternaria solani
		Tizón temprano
		Mancha negra de la
		   hoja
			Dürrfleckenkrankheit
			Dörrfleckenkrankheit
				Brûlure alternarienne
				Maladie des
				   taches brunes

False root-knot nematode
Nacobbus aberrans
		Falso nematodo del
		   nudo de la raíz
		Rosario
				Nematode cécidogene
				   de Cobb
				Nematode de gales
				   velues

Fusarium dry rots
Fusarium solani
F. roseum
	Fusarium storage rots
	Seed piece decay
		Pudriciones secas por
		   Fusarium
			Fusarium-Trockenfäule
			Fusarium-Weissfäule
				Fusariose
				Pourriture sèche
				   fusarienne

Fusarium wilts *Fusarium eumartii* *F. oxysporum* *F. avanaceum* *F. solani*		Marchitez por Fusarium	Fusarium-Welk	Flétrissure fusarienne
Gangrene *Phoma exigua*		Gangrena Cancro Pudrición de la raíz	Phoma-Stengelbraune Phoma-Trockenfäule Phoma-Knollenfäule	Gangrene Pourriture phoméenne
Gray mold *Botrytis cinerea*		Pudrición gris Moho gris	Grauschimmel Pustelfäule	Pourriture grise Moisissure grise
Hail injury		Daños provocados por granizo	Hagelschäden	Dégâts de grêle
Hair sprout	Spindle sprout	Brotes filiformes	Fadenkeimigkeit	
High temperature field injury		Daños a la planta por alta temperatura		
Hollow heart		Corazón vacío	Hohlherzigkeit	Coeur creux

Internal heat necrosis	Internal brown spot
	Internal rust spot
	Internal spotting
		Necrosis interna de
		   los tubérculos
			Eisenfleckigkeit
			Braunherzigkeit
			Kringerigheid
			   (Dutch)
				Taches de rouille

Internal sprouting	Ingrown sprouts	Brotamiento interno	Innerer Keimdurchwuchs	Germination introrse
Late blight *Phytophthora infestans*		Tizón tardío Hielo Gota Rancha	Krautfäule Knollenfäule Braunfäule	Mildiou
Leak *Pythium* spp.	Watery wound rot	Pudrición acuosa Gotera	Wassrige Wundfaule	Pourriture aqueuse

Lesion nematodes
Pratylenchus spp.
		Nematodo de la lesión
		   radicular de racines
				Nematode des lesions
				   de racines

Lightning injury		Daños provocados por relámpagos		

Low temperature
foliage injury
		Daños en el follaje por
		   baja temperatura
		Heladas
			Frostschäden an
			   der Pflanze
				Dégâts de froid

(continued on next page)

Common Name Causal Factor	Other Names	Spanish	German	French
Low temperature tuber injury		Daños en el tubérculo por baja temperatura	Frostschäden an der Knolle Kalteschaden an den Knollen	
Nonvirus leafroll		Enrollamiento no viral	Physiologisches Blattrollen	
Nutrient imbalance		Desbalance nutricional		
Phoma leaf spot *Phoma andina*	Black blight	Tizón foliar Tizón negro		
Pink eye *Pseudomonas fluorescens*	Red xylem disease Bruise infection Brown eye	Ojo rosado		Rosissement des yeux
Pink rot *Phytophthora erythroseptica* *Phytophthora* spp.	Watery rot Wilt	Pudrición rosada Podredumbre rosada	Rotfäule Rosafäule	Pourriture rose Pourriture humide
Pleospora herbarum		Mancha foliar por Pleospora		
Potato aucuba mosaic PAMV	Pseudo net necrosis Tuber blotch Viruses F and G	Mosaico aucuba Mosaico necrótico	Akuba Mosaik Aucubamosaik	Mosaique d'auchuba
Potato cyst nematodes *Globodera* spp. (*Heterodera* spp.)	Golden nematodes Potato root eelworm	Nematodo del quiste Nematodo dorado	Gelber Kartoffel nematode Kartoffelzystenälchen	Maladie de la pomme de terre Nematode sur pomme de terre
Potato leafroll virus PLRV	Potato phloem necrosis Tuber net necrosis	Enrollamiento Enrollado Enanismo amarillo	Blattrollkrankheit Blattrollvirus Knollennetznekrose	Enroulement
Potato mop-top PMTV	Mop-head Yellow mottling virus	Mop-top de la papa		
Potato rot nematode *Ditylenchus destructor*		Nematodo de la pudrición de la papa Pudrición seca de la papa	Kartoffelkrätzealchen Nematodenfäule der Kartoffel	Pourriture du tubercule Nematode de la pourriture du tubercule de pomme de terre
Potato spindle tuber viroid PSTV	Unmottled curly dwarf Tomato bunch top Gothic	Tubérculo ahusado	Spindelknollen- krankheit	Tubercules en fuseau Tubercules fusiforme
Potato virus A PVA	Mild mosaic	Mosaico suave	Rauhmosaik	Frisolée mosaique
Potato virus M PVM	Potato leafrolling mosaic Interveinal mosaic Paracrinkle Potato viruses E and K	Mosaico crespo	Rollmosaik	
Potato virus S PVS				
Potato virus T PVT				
Potato virus X PVX	Potato latent Potato mild mosaic Potato simple mosaic Healthy potato virus Potato viruses B and D	Mosaico latente Mosaico leve	Kartoffel X-Mosaik Leichtes Mosaik	Mosaique légère

(continued on next page)

Common Name / Causal Factor	Other Names	Spanish	German	French
Potato virus Y PVY	Rugose mosaic Streak Leafdrop streak Stipple streak Potato virus C	Mosaico rugoso Mosaico severo	Strichelkrankheit	Bigarrure Frisolée
Potato yellow dwarf PYDV		Enanismo amarillo	Gelbzwergigkeit	
Potato yellow vein PYVV	Vein yellowing virus	Amarillamiento de las nervaduras		
Powdery mildew *Erysiphe* *cichoracearum*		Oidiosis Mildiu pulverulento	Echter Mehltau	Oïdium
Powdery scab *Spongospora subterranea*	Corky scab	Roña Polvosa Sarna polvosa	Pulverschorf Kartoffelräude Räude	Gale poudreuse Gale spongieuse
Psyllid yellows		Amarillamientos por psíllidos		
Rhizoctonia canker *Rhizoctonia solani*	Black scurf	Rhizoctoniasis Costra negra	Wurzeltöterkrankheit Kartoffelpocken Pockenkrankheit	Rhizoctone brune
Rhizopus soft rot *Rhizopus* spp.		Pudrición blanda por Rhizopus		
Ring rot *Corynebacterium* *sepedonicum*	Bacterial ring rot	Pudrición anular	Bakterienringfäule Ringröte Bakterielle Schleimfäule	Flétrissement bactérien
Root-knot nematodes *Meloidogyne* spp.		Nematodo del nudo de la raíz	Wurzelgallenälchen	Nodosite des racines
Rosellinia black rot *Rosellinia* spp.	Torbo	Torbo Mortaja Lanosa		
Second growth Jelly end rot		Crecimiento secundario Pudricion apical gelatinosa	Glasigkeit Durchwuchs Zwiewuchs Kindelbildung	Anomalie de croissance Aspect vitreux
Secondary tubers	No-top Little potatoes	Tubérculos secundarios	Knöllchensucht	Boulage Couveuse
Septoria leaf spot *Septoria lycopersici*		Mancha anular de la hoja		
Silver scurf *Helminthosporium solani*		Costra plateada Mancha plateada Caspa plateada	Silberschorf Silberflecken	Gale argentee Tache argentee
Skin spot *Oospora pustulans*		Mancha de la cáscara	Tüpfelfleckenkrankheit	Tache de la pelure Moucheture du tubercule
Smut *Thecaphora solani*	Thecaphora smut	Carbón Buba		
Stem rot *Sclerotium rolfsii*	Southern blight	Pudrición basal	Sklerotium-Knollenfäule	
Stem-end browning		Bronceado de la base de los tubérculos		Brunissement du talon
Stemphylium consortiale				Tache stemphylienne
Stubby-root nematodes *Trichodorus primitivus* *Paratrichodorus* spp.		Nematodes de la atrofia radicular		
Sugar beet curly top BCTV	Green dwarf	Punta crespa		

(continued on next page)

Common Name Causal Factor	Other Names	Spanish	German	French
Surface abrasions		Peladura		
Tobacco mosaic TMV		Mosaico del tabaco		
Tobacco necrosis TNV	Potato ABC disease	Virus de la necrosis del tabaco	Tabaknekrosevirus	
Tobacco rattle TRV	Stem mottle Spraing Corky ringspot		Ratel-Virus Propfenbildung Profenkrankheit Stengelbuntkrankheit Stengelbunt Tabakmauchevirus	Tacheture de la tige
Tobacco ringspot TRSV	Andean potato calico			
Tomato black ring TBRV	Bouquet Pseudo-aucuba	Bouquet Pseudo-aucuba	Bukettkrankheit Bukettvirus Gelbfleckigkeit	Bouquet
Tomato spotted wilt TSWV	Spotted wilt	Marchitez apical Necrosis de los brotes Necrose do topo (Portuguese) Viracabeça (Portuguese)	Bronzefleckenkrankheit	
Tuber cracks		Agrietadura de los tubérculos	Rissigkeit Wachstumrisse	Craquelement
Tuber greening Sunscald	Sun-green	Verdeamiento Escaldadura	Grünverfärbung der Knollen	Verdissement Insolation
Ulocladium blight *Ulocladium atrum*		Kasahui		
Verticillium wilt *Verticillium albo-atrum* *V. dahliae*		Marchitez por Verticillium Verticilosis	Welkekrankheiten Wirtelpilz-Welkekrankheit Verticillium-Welke	Maladie du jaune Flétrissure verticilliene Verticilliose
Violet root rot *Rhizoctonia* *crocorum*		Pudrición radicular violeta	Violetter Wurzeltöter Violette Wurzelfäule	Rhizoctone violet
Wart *Synchytrium* *endobioticum*	Black wart	Verruga Roña negra	Kartoffelkrebs	Gale verruqueuse Tumeur verruqueuse Gale noire
White mold *Sclerotinia* *sclerotiorum* (also *S. minor*, *Sclerotinia* spp.)	Stalk break	Esclerotiniosis Moho blanco Pudrición dura	Sklerotinia- Stengelfäule	Porriture du collet Pourriture sclerotique
Wind injury		Daños provocados por el viento	Windschäden	Dégâts de vent
Witches' broom Mycoplasma	Northern stolbur Dwarf shrub virosis	Escoba de brujas	Hexenbesenkrankheit	Balai de sorcière

Glossary

abraded—rubbed or worn away, especially by friction; eroded

acervulus (pl. acervuli)—saucer-shaped or cushionlike fungus fruiting body bearing conidiophores, conidia, and sometimes setae

acre—unit of land area 43,560 ft^2 (0.40469 hectare; 4,046.87 m^2)

acute—developing suddenly; severe, e.g., symptoms of disease; or less than 90°, describing an angle

adjuvant—a substance added to a medicinal to aid its action

adventitious—arising not at its usual site; e.g., roots originating from stems, tubers, or leaves

aerial tuber—a tuber developing in the axils of leaves aboveground on potato stems

aerobic—requiring the presence of elemental oxygen for survival

agar—solidifying component of microbial culture media derived from certain marine, red algae

akaryotic—describing a cell without well-differentiated nucleus

amorphous—lacking a definite form or shape

AMV—alfalfa mosaic virus

anaerobic—living and surviving in absence of elemental oxygen

anastomoses (sing. anastomosis)—interconnections between branches of the same or different hyphae (or other structures) to make a network

antheridium (pl. antheridia)—fungus structure producing male gametes (male gametangium)

anthocyanin—blue, purple, red, or pink water-soluble flavenoid pigment in cell sap

antigen (adj. antigenic)—a foreign chemical, usually a protein, that induces antibody formation when injected into an animal body

antiserum (pl. antisera)—serum containing antibodies

apex (pl. apices, adj. apical)—tip of root or shoot containing the apical meristem

aphid—a small, sucking, homopterous insect living on plant juices and capable of transmitting viruses

APLV—andean potato latent virus

APMV—andean potato mottle virus

apothecium (pl. apothecia)—open cuplike or saucerlike, ascus-bearing fungus fruiting body

appressorium (pl. appressoria)—swelling on a fungus germ tube or hypha, especially for attachment to a host in an early stage of penetration

ascospore—spore formed within an ascus

ascus (pl. asci)—saclike cell in which ascospores (typically eight) are produced

ash—the solid, noncombustible residue left after burning

attenuated—reduced in virulence

aucuba—bright yellow mosaic leaf variegation of genetic or virus origin

avirulent—nonpathogenic

bacilliform—a blunt, thick rod shape, rounded on the ends; bacillus-shaped

bacillus—type of bacterium, rod-shaped with rounded ends

bacterium (pl. bacteria)—typically, a single-celled microorganism lacking chlorophyll and increasing by simple cell division

bar—metric unit of pressure, 1 bar = 0.987 atmosphere pressure, 10^6 dynes/cm^2

basidium (pl. basidia, adj. basidial)—a short, club-shaped fungus cell on which basidiospores are produced

BCTV—sugar beet curly top virus

bentonite—an absorptive and colloidal clay used especially as a carrier of chemicals (insecticides) or as ribonuclease inhibitor in virus extraction

bi- (prefix)—two

bicollateral—vascular bundle having phloem both outside and inside the xylem

binary fission—division of a cell into two daughter cells by simple division of the nucleoplasm and cytoplasm

biotype—subspecies of organisms morphologically similar but differing physiologically, particularly in ability to selectively parasitize plants with specific resistance

blight—a disease characterized by rapid and extensive death of plant foliage

broadcast application—fertilizer application by spreading or scattering within or on the soil surface

buffer—a substance capable in solution of keeping hydrogen-ion concentration constant and thereby avoiding rapid changes in acidity or basicity of a solution

°C—°Celsius (formerly °Centigrade), unit of temperature 0.01 between boiling and freezing points of water at standard pressure. °C = (°F − 32) 5/9 and °F = 9/5 (°C) + 32

calcareous—rich in calcium, often as carbonate, lime

callose—a carbohydrate component of plant cell walls often forming over sieve plates and in calcified cell walls

callus—a mass of parenchymatous cells formed over or around a wound

cambium—lateral meristematic layer of stems and roots, giving rise to secondary xylem, secondary phloem, and parenchyma and responsible for secondary growth

canker—necrotic, localized, diseased area

carbohydrate—various chemical compounds of carbon, hydrogen, and oxygen, such as sugars, starches, or cellulose

catenulate—formed in chains or in an end-to-end series

cellulose—a carbohydrate comprising the primary cell wall substance

certification scheme—a governmentally supervised procedure of seed propagation to insure high quality, varietal purity, and freedom from disease

chelate—relating to or having a ring structure that usually contains a metal ion held by coordination bonds

chimaera—plant with several tissues or tissue layers differing in genetic constitution

chip—in this text, a thin slice of potato tuber fried in deep fat

chlamydospore—thick-walled, asexual, resting spore formed by rounding up of a hyphal cell

chlorosis (adj. chlorotic)—abnormal plant color of light green or yellow due to incomplete formation or destruction of chlorophyll

chondriosomes—a generic term for small cytoplasmic structures including mitochondria

chromosome—elongate aggregate of genes formed within nuclei at certain stages of cell division

circulative—describing viruses that must accumulate within or pass through the lymphatic system of their insect vector before transmission to plants

cisterna (pl. cisternae)—a cavity within a cell enclosed by a membrane

clavate—club-shaped

cleistothecium (pl. cleistothecia)—closed, usually spherical, ascus-containing structure of powdery mildew fungi

clone—group of vegetatively (asexually) propagated plants derived from a single original plant or plant part

CMV—cucumber mosaic virus

coalesce—union of similar structures merging or growing together into a larger similar structure

coenocytic—multinucleate; e.g., a multinucleate plant body enclosed within a common wall or a fungus filament lacking cross walls

comovirus—a virus within the group to which cowpea mosaic virus belongs

conidiophore—specialized fungus hypha on which conidia (conidiospores) are produced

conidium (pl. conidia)—any asexually produced spore germinating by a germ tube

cortex (adj. cortical)—parenchymatous tissue between the epidermis and phloem in stems, tubers, and roots

cotyledon—seed leaf; primary embryonic leaf within the seed in which nutrient for the new plant is stored

cupulate—cuplike, cup-shaped

cuticle—water-repellent waxy covering (cutin) of epidermal cells of plant parts such as leaves, stems, or fruits; also the outer sheath or membrane of nematodes

cv. (cultivar)—a plant variety, a cultural selection

cwt—100 lb, 45.45 kg

cyst—a capsule around certain cells, as bacteria in a resting spore stage; also the egg-laden carcass of a female nematode

cystosori—a group of sporangia formed after division of a single protoplast

cyto- (prefix)—referring to cell

cytoplasm—substance of a cell body exclusive of the nucleus

damping off—rapid destruction and collapse of seedling plants near soil level due to cortical decay

decortication—loss of cortex due to rot

dehydrate—to reduce water content, to become dry

density gradient centrifugation—separation of components by centrifugation in a column of a solution of increasing density

desiccate—to dry out

diagnostic—a distinguishing characteristic important for identification of disease or other condition

dicotyledons (adj. dicotyledonous)—plants having two cotyledons (seed leaves), in contrast to monocotyledons (the grasses and cereals)

dilution end point—the point at which infectivity or other activity is lost due to dilution

diploid—having two sets of chromosomes (in potato 2n = 2x = 24)

distal—far or opposite from the end of attachment or origin

dolomitic limestone—limestone rich in magnesium carbonate, $CaMg(CO_3)_2$

dormant—resting, living but in a state of reduced activity

electrophoresis—movement of charged particles and macromolecular ions under the influence of an electric field

ELISA—enzyme-linked immunosorbent assay, an extremely sensitive serological test for virus or other antigens

elite seed—seed selected from basic stocks of known origin, varietal purity, and freedom from disease and protected from contamination by sanitation and isolation

encapsidated—enclosed as if in a capsule

encyst (n. encystment)—to become enclosed in a cyst, a capsule

endemic—native to or peculiar to a locality or region

endoplasmic—pertaining to the inner granular, relatively fluid part of the cytoplasm

enzyme—protein that catalyzes a specific biochemical reaction.

epicotyl—describing the portion of a plant embryo or seedling above the cotyledonary node

epidemiology—study of disease initiation, development, and spread, particularly as influenced by environment

epidermis (adj. epidermal)—outer layer of cells usually one cell thick on plant parts. On tubers, the epidermis is very short-lived

epinasty—downward curvature of leaf, leaf part, or stem due to rapid expansion of the upper surface

erose—having the margin irregularly notched as if gnawed

erumpent—breaking out or erupting through the surface

exudate—usually an ooze or slime discharged from a diseased plant part

facultative—capable of changing life style; e.g., from saprophytic to parasitic or the reverse

fallow—describing plant-free cultivated land kept free from a crop or weeds during the normal growing season

fasciated—malformed by growing together of plant structures, stems, or buds

filament (adj. filamentous)—thin, flexible, threadlike structure

filiform—threadlike

fixation—preservation of biological structures for microscopic examination by killing in suitable chemicals or physical conditions so as to avoid changes in structure

flaccid—wilted, lacking in turgor

flagellum (pl. flagella, adj. flagellar)—hairlike or whiplike appendage of bacterial cells or fungus zoospores providing movement

flocculation—aggregation into a loose fluffy mass

fructification—in fungi, a spore-bearing structure

fumigant—a vapor-active chemical used in the gaseous phase to kill or inhibit growth of microorganisms or other pests

fungicide—a substance killing fungi; sometimes broadly used also for substances inhibiting growth of fungi or spore germination

fungus (pl. fungi)—spore-producing plant lacking chlorophyll, often causing disease of higher plants

g—gram, a unit of metric weight, approximately 1/28 oz

galls—localized enlargements (overgrowths) on plants

gelatinous—resembling gelatin or jelly

gel-diffusion—a type of serological assay for virus identification

gemmation—in potato, successive production of tubers on a stolon in a beadlike manner

genetic—relating to heredity; describing heritable characteristics as influenced by germplasm

genotype—the entire genetic constitution of an organism

geotropic—plant growth directed toward the force of gravity; e.g., roots

germ tube—initial hyphal strand from a germinating fungus spore

germplasm—material capable of transmitting heritable characteristics sexually or asexually

giant cells—multinucleate cells formed by disintegration of cell walls; also called syncytia in nematode infections

glycoprotein—a conjugated protein in which the nonprotein group is carbohydrate

Gram stain—a stain for differentiating bacterial types

greening—development of chlorophyll in tubers after exposure to light

ha—hectare, 10,000 m^2 (2.47 acres)

haploid—having the single basic chromosome number as in most germ cells

haulms—plant stems or stalks, vines of potato

haustorium (pl. haustoria)—specialized fungus protuberance into a host cell, probably functioning in food absorption

herbaceous—nonwoody; e.g., a plant or plant part

herbicides—chemicals that kill herbaceous plants; also applied to those that limit growth of such plants

heterologous—different although apparently similar; e.g., the reaction between an antiserum and an antigen closely resembling but not identical to the antigen causing the production of antibody

hexaploid—having six sets of the basic number of chromosomes (in potato, 2n = 6x = 72)

histopathology—study of pathology of cells and tissues; microscopic changes characteristic of disease

homogeneous—similar in certain characteristics, such as in chemical nature or physical properties

host—plant that furnishes a medium suitable for development of a parasite

hyaline—colorless, transparent

hybrid—sexually produced offspring of parents differing genetically. In potato, further vegetative propagation may continue as a clone.

hydrated—having absorbed water

hydrolyzed—having undergone chemical decomposition involving splitting of a bond and addition of hydrogen and oxygen

hyperplasia—abnormal increase in the number of cells, resulting in formation of galls or tumors

hypersensitive—extremely or excessively sensitive; having a type of resistance resulting from extreme sensitivity to a disease

hypertrophy—abnormal increase in the size of cells, resulting in formation of galls or tumors

hypha (pl. hyphae)—tubular filament of a fungus

hyphal fusion—joining of fungal hyphae, usually with some exchange of cell contents

hypocotyl—the part of a plant embryo or seedling below the cotyledons

icosahedral—describing a regular polyhedron with 20 equilateral-triangular faces

immunity—high resistance against a disease, exemption from infection; or in an animal, having developed antibodies against a foreign substance (usually a protein)

immunogenic—producing immunity, usually describing a protein (antigen) capable of causing antibody formation when injected into an animal

in vitro—in an artificial environment, usually outside the living body

in vivo—in a living body

incipient—early in development (of a disease or condition)

inclusion—nonprotoplasmic structure inside a cell

indicator host—plant that responds specifically to a particular infection, used to detect a disease or to identify the pathogen

indigenous—native

infection—entrance and subsequent multiplication of a microorganism in a plant

infection court—site in or on host plant where infection can occur

infection propagules—infectious units of inoculum

inoculum—parts of a pathogen capable of infecting a host

intercalary—situated between existing layers or plant parts

intercellular—between cells

intercostal—between veins, interveinal

internode—portion of the stem between joints or leaf attachments

interveinal—between veins

intracellular—within cells

irradiation—exposure to radiant energy of various types

isometric—equally long, as a virus particle with all axes of equal length (essentially spherical)

kg—kilogram, 1,000 g (2.2 lb)

labile—unstable

lamina (pl. laminae)—a layer; the broad expanded part of a leaf

larva—juvenile stage of certain animals (e.g., nematodes and aphids) occurring between the embryo and the adult

latent—present but invisible or inactive

lateral buds—buds formed on stems at the axils of leaves

latex—rubberlike

leaching—removal of a chemical through solubility, usually in water

legumes—plants belonging to the Leguminosae, including beans, peas, alfalfa, and clover

lenticel—natural opening in surface of leaf, stem, or tuber permitting gas exchange

lenticular—lens-shaped (convex on both faces)

lesion—distinct diseased area

leucoplast—colorless plastid

lipid—generic term for oils, fats, waxes, and related products found in living tissues

local lesion host—a host (usually of a virus) responding by lesions at the site of infection

locule (adj. locular)—a cavity, especially one in a fungus stroma

lysogeny—dissolution; cell destruction by dissolution

macerate—to cause to become softened and desintegrated as by steeping or soaking in fluid

marl—a type of soil, rich in lime, formed in the bottom of a lake or swamp

mechanical injury—injury of a plant part by abrasion, mutilation, or wounding

medullary—of or relating to the pith of a plant

melanin—dark to black pigment

meristem—plant tissue functioning principally in cell division

meristem culture—aseptic culture of a plant or plant part from a portion of the meristem

mesophyll—central, internal, nonvascular tissue of a leaf, consisting of the palisade and spongy mesophyll

microbial—pertaining to or relating to microbes or microorganisms

microorganism—an organism of microscopic size

microprecipitin test or precipitin test—a type of serological test for virus

microsclerotia—very small sclerotia

microtubules—any of the minute cylindrical structures of a cell that are widely distributed in protoplasm and are made up of longitudinal fibrils

mildew—superficial (surface) fungus growth

mitochondria—various long or round cellular organelles that are found outside the nucleus of a cell, produce energy for the cell through respiration, and are rich in fats, proteins, and enzymes

MLO—mycoplasmalike organisms

mm—millimeter, 1/1000 of a meter, approximately 1/25 in.

μ**m**—micron or micrometer, 10^{-6} m, approximately 1/25,000 in.

molecular weight—the weight of a molecule expressed as the sum of the atomic weights of its constituent atoms

molecule—the smallest particle of a substance composed of one or more atoms that retains the properties of the substance

monocotyledons (adj. monocotyledonous)—plants (including the grasses) with one seed leaf

monogenic resistance—resistance determined by a single gene

morphology—study of form and structure

mosaic—disease symptom usually of a virus; nonuniform foliage coloration; a more or less distinct intermingling of normal, light green, or yellowish colored patches; a mottle

motile—exhibiting or capable of movement

mottle—disease symptom comprised of light and dark areas, an irregular pattern on a leaf

muck soil—soil similar to peat soil, often having a lower percentage of organic materials

multinucleate—having more than one nucleus enclosed within a cell wall

muriform—having cells like bricks in a wall with both longitudinal and transverse septa

mutation—heritable genetic change in a cell

mycelium—hyphae compromising the thallus or body of a fungus

mycoplasma (mycoplasm)—procaryotic organism, smaller than bacteria and larger than viruses, without rigid cell walls and varying in shape, reproducing by budding or fission

necrosis (adj. necrotic)—death of plant cells or plant parts, usually accompanied by darkening or discoloration; a symptom of disease

nematicide—chemical agent that kills nematodes

nematode—threadlike round worms of the order Nematoda, usually soilborne, of which a number of microscopic size attack potatoes

net necrosis—necrosis of phloem tissues within tubers causing a netlike pattern of internal discoloration

nm—nanometer, 10^{-9}m, 0.001 μm

node—joint in a stem, also the eye of tuber at which leaves and axillary buds are formed

nonpersistent—short-lived; said of viruses that are infectious for only short periods when transferred in or on insect mouthparts

nonseptate—describing fungus filaments without cross walls

nymph—juvenile stage of insect with incomplete metamorphosis but superficially resembling the adult

obovate—egg-shaped with wide end outward

obovoid—egg-shaped with narrow end outward

omnivorous—feeding on substances of both animal and vegetable origin

oogonium (pl. oogonia)—the female egg cell of oomycete fungi

oospore—thick-walled, sexually derived resting spore of phycomyceteous fungi

organelle—delimited membranous structure within a cell having a specialized function

ostiole—pore; opening in a perithecium or pycnidium

ozone—O_3, a photochemical oxidant air pollutant

palisade—a layer or layers of columnar cells rich in chloroplasts present beneath the upper epidermis of plant leaves

PAMV—potato aucuba mosaic virus

papillum (pl. papilla)—small, round or nipplelike projection

paracrinkle—a symptom of mild crinkle in virus infections

paragynous—having the antheridium at the side of the oogonium

paraphyses—hairlike cells within a fungus fruiting structure

parasite—organism that lives with, in, or on another organism (host) to its own advantage and to the disadvantage of the host

parenchyma—soft tissue of living plant cells with undifferentiated, thin, cellulose walls

pathogen (adj. pathogenic)—the causal agent of a disease

peat soil—a soil type, high in organic materials consisting of partially decayed, moisture-absorbing plant materials, formed in bogs or swamps

pectolytic—enzyme capable of dissolving pectin (the substance that normally cements plant cells together)

pedicel—stalklike structure

pentaploid—having five sets of chromosomes (in potato, $2n = 5x = 60$)

peptone—any of various water-soluble products following partial hydrolysis of proteins

perennial—a plant naturally persisting vegetatively for more than one year or growing season

periclinal chimaeras—plants with inner tissues genetically different from outer tissues

peridial—referring to the outer envelope of the sporophore of many fungi

pericycle—a thin layer of parenchymatous or sclerenchymatous cells that surrounds the stele in most vascular plants

permeability (adj. permeable)—the quality or condition allowing a fluid or substance in a fluid to pass or diffuse through a membrane

persistent—describing a relationship between virus and vector characterized by a lapse of several hours between acquisition and first transmission and the continuation of virus transmission for many days following removal of the insect from the virus source

petiole—stalklike portion of a leaf attached to the stem and supporting the lamina

pH—measurement of acidity or basicity; pH 7 being neutral, values below being acid, and those above being basic (alkaline)

phenol (adj. phenolic)—a toxic acidic compound, C_6H_5OH, used as a disinfectant or protein denaturant

phenolase—an enzyme capable of degrading phenolic compounds

phloem—vascular tissue consisting usually of sieve tubes, companion cells, and parenchyma that conducts elaborated food materials

photochemical oxidants—highly reactive compounds formed by action of sunlight on less toxic precursors

photodegredation—degradation due to light, usually sunlight

phyllody—change of a plant organ into a foliage leaf

phytotoxic—harmful to plants; usually describing a chemical

pigmentation—coloration

pinnate—describing leaves having similar parts arranged on opposite sides of the axis

pitch—in a filamentous virus particle, the axial distance between adjacent turns of a row of capsids

pith—loose, spongy tissue in the center of certain stems

plasmodium (pl. plasmodia)—naked mass of protoplasm without cell walls containing nuclei and cytoplasm, usually of a fungus

plastid—any of various cytoplasmic organelles (chloroplasts, leucoplasts, etc.) that serve in many cases as centers of special metabolic activities

pleomorphic—with various shapes; of nonconstant form

PLRV—potato leaf roll virus

PMTV—potato mop-top virus

podzol—type of light colored, relatively infertile soil of cool, coniferous forests poor in lime and iron

pollen—male sex cells produced by anthers of flowering plants

polymerize—to subject to or undergo a chemical reaction in which two or more similar molecules combine to form larger molecules of repeating structural units

polyploidy—state of having more than two chromosome sets

polysaccharide—a carbohydrate that can be decomposed by hydrolysis into two or more molecules of monosaccharides

ppm—parts per million

primary inoculum—inoculum, usually from an overwintering source, that initiates disease in the field, rather than that which spreads disease during the season

primary symptom—the symptom produced soon after infection, in contrast to a secondary symptom, which follows more complete invasion

primordium (adj. primordial)—the rudimentary or initiating portion from which a plant part is formed

progeny—descendants, offspring

propagule—any part of an organism capable of independent growth

protein—any of numerous naturally occurring, complex combinations of amino acids, which are essential constituents of all living cells

protoxylem—the first-formed xylem, with annular, spiral, or scalariform wall thickenings

pseudosclerotia—sclerotialike structures

PSTV—potato spindle tuber viroid

psyllids—jumping plant lice of the family Psyllidae

punctate—dotlike, marked with dots or tiny spots

pustule—blisterlike; small erumpent spot, spore mass, or sorus

PVA—potato virus A

PVM—potato virus M

PVS—potato virus S

PVT—potato virus T

PVX—potato virus X

PVY—potato virus Y

pycnidiospores—spores (conidia) produced in a pycnidium

pycnidium (pl. pycnidia)—asexual, globose or flask-shaped fruiting body of fungi producing conidia

PYDV—potato yellow dwarf virus

PYVV—potato yellow vein virus

quinones—any of various (usually yellow, orange, or red) quinonoid compounds, including several that are biologically important as coenzymes, hydrogen acceptors, or vitamins

race—biotype

reducing sugars—sugars with free carbonyl groups such as fructose, formed from hydrolysis of complex sugars

resistance (adj. resistant)—property of hosts that prevents or impedes infection or disease development

resorption—the action of absorbing again a substance previously differentiated

respiration—enzymatic reactions within a living organism utilizing O_2 and releasing CO_2, usually for production of energy

resting spore—temporarily dormant spore, usually thick-walled, and capable of surviving adverse environments

reticulum (adj. reticular, reticulate)—netlike or weblike structure

rh—relative humidity

rhizome—horizontal underground stem of more than one year's growth, possessing buds, nodes, and usually scalelike leaves

rhizomorph—fungus mycelium arranged in strands, rootlike in appearance

rhizosphere—microenvironment in soil near to and influenced by plant roots

ribonucleic acid—any of a number of nucleic acids containing ribose, uracil, guanine, cytosine, and adenine and associated with control of cellular chemical activity; the nucleic acid type of most plant viruses

RNA—ribonucleic acid

rogue (noun)—diseased or abnormal plant; **(verb)**—to remove rogues during their growth

root cap—protective cap covering apical meristem at root tip

rosario—arranged as beads on a string (Spanish for rosary)

rugose mosaic—severe mosaic accompanied by deformation such as leaf crinkling, curling, or roughening of leaf surface

saprophyte—nonpathogenic plant that obtains nourishment from the products of organic breakdown and decay

scald—a necrotic condition of tissue, usually bleached in color, with appearance of having been exposed to high temperature

sclerotia—drought-resistant or heat-resistant form of fungus structure, usually with thick, hard cell walls permitting survival over adverse environments

second growth—resumption of growth after normal growth has ceased

secondary organism—organism that multiplies in already diseased tissue; not the primary pathogen

secondary rot—rot caused by a secondary organism

secondary symptom—symptom of virus infection appearing after first (primary) symptoms; in potato, a symptom often from infection borne by seed-tubers

seed tubers—tubers or tuber parts planted as seed for asexual propagation of potato

senesce (n. senescence)—to decline with maturity or age; often hastened by stress from environment or disease

septum (pl. septa)—cross wall

serological method—several types of tests for identifying viruses by using an antiserum that reacts specifically with a given virus protein

serum—colorless, liquid component of blood used in serological tests for viruses

seta (pl. setae)—bristlelike fungus structure

sieve tube—a tube consisting of an end-to-end series of thin-walled cells in the phloem, with ends (sieve plates) perforated and thickened, functioning chiefly in translocation of organic solutes

solanine—a potentially toxic glycoalkaloid present in plants of the Solanaceae, including potato

somatic—relating to the body, especially body cells as distinguished from germplasm

somatic aberration—mutation or abnormality in a somatic cell and its progeny

sorus (pl. sori)—a group of spores that is formed within plant tissue and that may erupt through the surface

sp. (singular, pl. spp.)—species

specific gravity—in potatoes, dry matter content of tubers expressed as weight per unit of volume; used as an indication of starch content

sporangium (pl. sporangia)—a type of fungus structure producing asexual spores, usually zoospores

sporangiophore—a sporangium-bearing body of a fungus

spore—reproductive body of fungi and other lower plants, containing one or more cells; a bacterial cell modified to survive adverse environment

sporiferous—bearing or producing spores

sporophore—a spore-bearing body in fungi

sporulating—producing and often liberating spores

ssp.—subspecies

sterigma (pl. sterigmata)—small, usually pointed protuberance on which basidiospores are borne

sterile—free from contaminant organisms; incapable of propagation; infertile

stolon—type of underground stem on the tip of which, in potato, tubers are formed

stomate (pl. stomata)—opening in the epidermis of a plant part surrounded by guard cells and leading to an intercellular space through which gases diffuse

strain—a selection of an organism with peculiar characteristics; race; biotype

stylet—slender tubular mouthparts in plant-parasitic nematodes or aphids

subculture—a culture (e.g., of bacteria or fungi) derived from another culture

suberin—waxy, water-impervious substance associated with corky tissue deposited in or on plant cells

suberized—infiltration of cell walls with suberin

suboxidation—deficiency of oxygen that impairs normal respiration

substrate—the substance on which an organism lives or from which it obtains nutrients; chemical substance acted upon, often by an enzyme

succulent (noun or adj.)—plant, or plant part with tender, juicy, or watery tissues

supercooled—describing liquid cooled below its freezing point without formation of ice crystals

surfactant—a surface-active substance modifying surface tension of liquids and their ability to wet other materials

susceptible—lacking resistance; prone to infection

syn.—synonym

syncytium (pl. syncytia)—multinucleate mass of protoplasm resulting from continued nuclear division or cell wall breakdown and subsequent fusion of protoplasts and surrounded by a common cell wall; also called giant cells in nematode infections

TBRV—tomato black ring virus

tensile strength—the greatest lengthwise stress a substance can bear without tearing apart

tetraploid—having four sets of chromosomes (in potato, $2n - 4x - 48$)

thermal inactivation temperature—the temperature at which infectious entities are inactivated usually within a given time period

tilth—state of soil aggregation or consistency; good tilth implying porous, friable texture

TMV—tobacco mosaic virus

TNV—tobacco necrosis virus

tolerance—capacity of a plant or crop to sustain disease or endure adverse environment without serious damage or injury

toxic—capable of causing injury to a living organism

toxin—poisonous substance of biological origin

translocate (n. translocation)—to transfer from one location to another in the plant body

translucent—permitting light to shine through as diffused light

transmit—to spread or transfer, as an infectious pathogen from plant to plant or from one plant generation to another

trichome—plant epidermal hair, of which several types exist

triploid—having three sets of chromosomes (in potato, $2n = 3x = 36$)

triturate—to grind as with a mortar and pestle

TRSV—tobacco ringspot virus

true seed—seed resulting from sexual fusion of gametes (in contrast to seed tubers of potato, which are produced asexually)

TRV—tobacco rattle virus

TSWV—tomato spotted wilt virus

tuber—short, thickened, fleshy underground stem, borne usually at the end of a stolon

tuber indexing—propagation of a plant from a tuber or a tuber part to determine presence of a tuberborne disease

tuberize—to form tubers

turgid—distension of cells or tissues due to water absorption

turgor pressure—internal pressure within plant cells. Lack of turgor causes plants to wilt.

tyrosinase—an enzyme widespread in plants and animals that catalyzes oxidation of tyrosine

tyrosine—a metabolically important phenolic amino acid

uninucleate—having one nucleus

vascular—pertaining to conductive (xylem and phloem) tissues

vascular ring—circular arrangement of vascular strands within a stem or tuber

vascular strand—group of elongated cells including xylem, phloem, and parenchymatous tissue, providing for movement of water and solutes and for mechanical support

vector—agent that transmits inoculum and is capable of disseminating disease

vegetative—referring to somatic or asexual parts of the plant not involved in sexual reproduction

viroid—the smallest known infectious agent consisting of nucleic acid and lacking the usual protein coat of viruses

virulent—pathogenic; having capacity for causing disease

viruliferous—virus-carrying; usually an insect or nematode

virus—an infective particle smaller than a bacterium containing protein and nucleic acid and capable of multiplying within plant or animal cells

volunteer plant—a potato plant growing from an unharvested tuber as a weed

xylem—complex woody tissue consisting of vessels, tracheids, fibers, and parenchyma that transports water and solutes and may serve also for mechanical support

zonate (n. zonation)—marked with stripes or lines more or less parallel to the edge of the lesion

zoosporangium—a spore case or sporangium-bearing zoospores

zoospore—fungus spore with flagella capable of locomotion in water

zygote—sexually produced cell formed by the union of two gametes

Index

Abrasions, tuber surfaces, 14
Aceratagallia sanguinolenta, potato yellow dwarf virus vector, 82
Actinomyces scabies, see Streptomyces scabies
Acyrthosiphon spp., 102
Aecidium cantensis, rust from, 65
Agallia quadripunctata, potato yellow dwarf virus vector, 82
Agrobacterium sp., in healthy tubers, 33
Air pollution, injury from, 20
Alfalfa mosaic virus, 82
  symptoms, 83
  control, 83
Alternaria
  alternata, 44
  solani, 43
  tenuis, 44
Aluminum, toxicity from, 24
Andean potato latent virus
  symptoms, 78
  control, 78
Andean potato mottle virus
  symptoms, 77
  control, 77
Angiosorus solani, smut caused by, 64
Aphids
  description, 101
  table, key to wingless forms, 102
Aphis
  fabae, 102
  frangulae, potato virus M vector, 75
  gossypii-frangulae, 102
  nasturtii, potato virus M vector, 75
Apical leafroll, 91
Armillaria mellea, tuber rot from, 66
Aster yellows, symptoms, 91
Aulacorthum solani, 102
Avenaceum wilt, *see* Fusarium wilts

Bacillus spp.
  soft rot caused by, 28
  megaterium, in healthy tubers, 33
Bacterial wilt, symptoms, 29
Belanolaimus longicaudatus, 94
Black dot
  Colletotrichum cause, 55
  silver scurf similarity, 55
Black scurf
  cause, 53
  control, 54
  symptoms, 52
Blackheart
  cause and symptoms, 10
  temperature relation, 9
Blackleg, symptoms, 27
Blackspot, potassium deficiency predisposing to, 23
Blight, *see* Early blight or Late blight
Blitecast, late blight epidemic forecasting, 42
Blue stem, 91
Boron deficiency, symptoms, 24

Botrytis cinerea
  gray mold caused by, 48
  ozone predisposing to infection by, 21
Brown rot
  ring rot compared to, 29
  symptoms, 29
Browning
  internal mahagony, temperature relation, 9
  stem-end, symptoms, 22
Bunch top, 91

Calcium deficiency, symptoms, 23, 24
Catechol test, use of, 15
Cavariella pastinacae, 102
Cercospora
  concors, leaf blotch caused by, 47
  solani-tuberosi, leaf blotch caused by, 47
Cercospora leaf blotch
  cause and symptoms, 47
  control, 47
Certification, seed potato, 104
Certified seed potato, fraudulent use, 104
Chaetomium spp., leaf spots from, 66
Charcoal rot
  cause and symptoms, 56
  control, 57
Chemical injury, symptoms, 21
Chimaeras, periclinal, 7
Choanephora blight, 48
Choanephora cucurbitarum, 48
Circulifer tenellus, sugar beet curly top virus vector, 90
Clonostachys araucariae, tuber rot from, 66
Clostridium spp., soft rot relation, 28
Colletotrichum
  atramentarium, 55
  coccodes, 55
Common scab, *see* Scab
Corticium, see Rhizoctonia
Corynebacterium sepedonicum, ring rot caused by, 31, 32
Cucumber mosaic virus, 79
Cylindrocarpon tonkinesis, dry rot from, 66

Deforming mosaic, symptoms, 87
Didymella sp., leaf spot from, 66
Ditylenchus spp., 94
  destructor, 100
Dwarf shrub virosis, 92

Early blight
  cause and symptoms, 43
  control, 44
Eisenfleckigkeit, 11
Enanismo amarillo, 68
Epitrix sp.
  Andean potato latent virus vector, 78
  hirtipennis, tobacco ringspot vector, 85
Erwinia
  carotovora, blackleg caused by, 27
  chrysanthemi, blackleg caused by, 28

Erysiphe cichoracearum, powdery mildew caused by, 43
Ethylene, coiled sprout role, 18
Eumartii wilt, *see* Fusarium wilts

False root-knot nematodes, 98
Flavobacterium sp.
  in healthy tubers, 33
  pectinovorum, soft rot from, 28
Foliage, temperature effect on, 9
Foundation seed, 104
Frankliniella spp., tomato spotted wilt virus vector, 87
Frost injury
  foliage, 9
  tubers, 9
Fungicide application, principles, 66
Fusarium
  oxysporum f. sp. *tuberosi,* 61
  roseum
    'Avenaceum', 59, 61
    'Sambucinum', 59
  solani, 61
    'Coeruleum', 59
    f. sp. *eumartii,* 61
Fusarium dry rots
  cause, 59
  control, 59
  symptoms, 58
Fusarium wilts
  cause, 61
  control, 61
  symptoms, 60

Gangrene
  cause and symptoms, 57
  control, 58
Gemmation, 12
Genetic abnormalities, 7
Giant hill, genetic abnormality, 7
Gilmaniella humicola, tuber rot from, 66
Globodera spp., 94
  pallida, 95
  rostochiensis, 95
Glomus fasciculatus, 66
Gram stain, recipe, 31
Gray mold
  cause and symptoms, 48
  control, 48

Hail injury, 19
Haywire, 91
Heat necrosis, internal, symptoms, 11
Helicobasidium purpureum, violet root rot caused by, 54
Helicotylenchus spp., 94
Helminthosporium solani, silver scurf caused by, 55
Heterodera, see Globodera
Heterosporium sp., on leaves and tubers, 66
Hexatylus vigissi, 94

Hollow heart, cause and symptoms, 13
Hyadaphis erysimi, 102
Hyalesthes obsoletus, stolbur vector, 93
Hypochnus, see Rhizoctonia

Igel-Lange test, tuber indexing, 69
Importance of potato, worldwide, 1
Insect toxins, 93
Internal net necrosis, 68

Late blight
   cause and symptoms, 40
   control, 42
Latebreaking virus, 91
Leafroll
   nonvirus, 18
   virus, *see* Potato leafroll virus
Leaf spots, 66
Leak
   cause and symptoms, 38
   control, 39
Lenticels, infection role, 4
Leptosphaerulina sp., leaf spot from, 66
Lesion nematodes, 99
Lightning injury, 19
Longidorus maximus, 94

Macrophomina phaseoli, charcoal rot cause, 56
Macrosiphoniella sanborni, 102
Macrosiphum euphorbiae, 102
   potato virus M vector, 75
Macrosporium solani, see Alternaria solani
Macrosteles fascifrons, aster yellows vector, 93
Magnesium deficiency, symptoms, 24
Manganese deficiency, symptoms, 26
Meloidogyne spp., 94, 97
Meloinema sp., 94
Micrococcus sp., in healthy tubers, 33
Mildew, *see* Powdery mildew
Moron, 91
Mutation, somatic, importance, 7
Mycelia sterilia, black rot caused by, 51
Mycoplasmas, description, 91
Mycorrhizae, 66
Mycovellosiella concors, leaf blotch caused by, 47
Myzus spp., 102
   ascolonicus, potato leafroll vector, 102
   persicae, 102
      potato leafroll vector, 68
      potato virus M vector, 75

Nacobbus aberrans, 94
   false root-knot cause, 98
Nasonovia lactucae, 102
National Plant Quarantine Act, 104
Nematicides, description, 101
Nematodes
   cyst, 95
   description, 93
   rot, 100
   table of, 94
Neoliturus tenellus, sugar beet curly top vector, 90
Neomyzus cucumflexus, 102
Neotylenchus abulbosus, 94
Net necrosis, low temperature causing, 8
Nitrogen, requirements and deficiency
   symptoms, 22, 23
Northern stolbur, 92

Olpidium brassicae, tobacco necrosis virus, 86
Oospora pustulans, see Polyscytalum pustulans
Ophiola falvopictus, witches' broom vector, 92
Oxygen requirements, tuber effect, 8
Oxysporum wilt, *see* Fusarium wilts
Ozone injury, symptoms, 20

Parastolbur, 91

Paratrichodorus spp., 94, 101
   tobacco rattle virus vector, 81
Paratrioza cockerelli, psyllid yellows cause, 93
Paratylenchus spp., 94
Pellicularia, see Rhizoctonia
Periconia sp., leaf spots from, 66
Peroxyacetyl nitrate, injury from, symptoms, 20
Phoma
   andina, leaf spot caused by, 47
   exigua, gangrene caused by, 57
Phoma leaf spot
   cause and symptoms, 47
   control, 47
Phosphorus, requirements and deficiency, 23
Phytophthora
   erythroseptica, pink rot from, 39
   infestans, late blight from, 40
Pink eye, symptoms, 32
Pink rot
   cause and symptoms, 39
   control, 40
Pleospora herbarum, symptoms, 46
Polyscytalum pustulans, skin spot caused by, 37
Potassium deficiency, symptoms, 23
Potato
   cultivated types, 2
   flowers, fruits, stems, roots, 2
   importance, worldwide, 1
   tubers, *see* Tuber
Potato aucuba mosaic virus
   control, 84
   symptoms, 84
Potato cyst nematodes
   cause, 95
   control, 96
   symptoms, 94
Potato leafroll, control, 69
Potato leafroll virus, symptoms, 68
Potato mop-top virus
   control, 80
   symptoms, 79
Potato rot nematodes
   control, 100
   symptoms, 100
Potato spindle tuber viroid
   control, 90
   symptoms, 89
Potato tuber blotch virus, 84
Potato virus A
   control, 72
   symptoms, 71
Potato virus G, 84
Potato virus M
   control, 75
   symptoms, 74
Potato virus S
   control, 76
   symptoms, 75
Potato virus T
   control, 77
   symptoms, 77
Potato virus X, 72
   control, 74
   symptoms, 73
Potato virus Y
   control, 71
   symptoms, 70
Potato yellow dwarf virus
   control, 82
   symptoms, 82
Potato yellow vein virus, symptoms, 86
Powdery mildew
   cause, 43
   control, 43
   symptoms, 42
Powdery scab, *see* Scab
Pratylenchus spp., 94
   control, 100
   symptoms from, 99

Pseudomonas sp.
   in healthy tubers, 33
   fluorescens, pink eye caused by, 32
   solanacearum, brown rot caused by, 30
Psyllid yellows, 93
Puccinia pittieriana, rust caused by, 65
Purple dwarf, 91
Purple top roll, 91
Purple top wilt, 91
Pythium spp., leak caused by, 8, 38

Rhizoctonia
   crocorum, 54
   solani, black scurf caused by, 53
Rhizoctonia canker
   cause, 53
   control, 54
   symptoms, 52
Rhizopus
   arrhizus, 52
   stolonifer, 52
Rhizopus soft rot
   cause and symptoms, 52
   control, 52
Rhopalosiphoninus spp., 102
Rhopalosiphum spp., 102
Rindite, 105
Ring rot
   brown rot compared to, 29
   cause and symptoms, 31
Root-knot nematodes
   control, 98
   symptoms, 97
Rosario, 98
Rosellinia spp., 51
Rosellinia black rot
   cause and symptoms, 51
   control, 52
Rotylenchulus spp., 94
Rotylenchus spp., 94
Rust
   common
      control, 65
      symptoms, 65
   deforming
      control, 66
      symptoms, 65

Scab
   common
      cause and symptoms, 33
      control, 34
   powdery
      cause and symptoms, 35
      control, 36
Scleracus flavopictus, witches' broom vector, 92
Scleroracus spp., witches' broom vectors, 92
Sclerotinia
   bataticola, 56
   minor, 49
   sclerotiorum, white mold caused by, 49
Sclerotium rolfsii, stem rot cause, 50
Seed improvement programs, 105
Seed treatment, tubers, 67
Septoria leaf spot
   cause and symptoms, 46
   control, 46
Septoria lycopersici
   control, 46
   symptoms, 46
Silver scurf, *Helminthosporium solani* causing, 54
Skin spot
   cause and symptoms, 37
   control, 38
Smut, *Thecaphora* causing, 63
Smynthurodes betae, 102
Solanum
   acaaule, 73

frost tolerance in, 9
andigena, 68
   spp., comparison, 2
chacoense, 71, 72
demissum, 71, 72
   late blight resistance in, 42
   species comparison, 2
   hybrids, diploids, triploids, tetraploids,
      pentaploids, hexaploids, 2
kurtzianum, 95
multidissectum, 95
phureja, spp. comparison, 2
rostratum, potato virus S host, 76
   spp., comparison of, 2
stenotomum, spp. comparison, 2
stoloniferum, 71, 72
tuberosum, 68, 73, 76, 95
   spp. comparison, 2
vernei, 95
Southern bacterial wilt
   symptoms, 29
   synonyms, 29
Spondylocladium atrovirens, 55
Spongospora subterranea
   powdery scab caused by, 35
   quarantine for, 104
Spraing, tobacco rattle virus causing, 80
Sprout
   coiled, cause and symptoms, 17
   hair, cause, 18
Sprouting, internal, in storage, 17
Stem rot
   cause and symptoms, 50
   control, 51
Stem streak necrosis, Mn toxicity causing, 26
Stem-end browning
   control, 22
   symptoms, 22
Stemphylium
   atrum, symptoms, 46
   botryosum, symptoms, 46
   consortiale, symptoms, 46
Stolbur, symptoms, 91
Streptomyces scabies, common scab caused
   by, 33
Stubby-root nematodes
   control, 101
   symptoms, 101
Sugar beet curly top virus, 90

Sulfur
   deficiency, symptoms, 24
   oxide, injury from, 21
Synchytrium endobioticum
   potato virus X vector, 73
   quarantine for, 104
   wart caused by, 36

Tall types, genetic abnormality, 7
Temperature
   foliage affected by, 9
   of soil, injury from, 10
   tubers affected by, 8
Thanetephorus, see Rhizoctonia
Thecaphora solani, smut caused by, 64
Thrips tabaci
   tobacco ringspot virus vector, 85
   tomato spotted wilt vector, 87
Tobacco black ring virus
   control, 86
   symptoms, 85
Tobacco mosaic virus, 79
Tobacco necrosis virus, symptoms, 86
Tobacco rattle virus
   control, 81
   symptoms, 80
Tobacco ringspot virus
   control, 85
   symptoms, 84
Tomato big bud, 91
Tomato black ring virus
   cause and symptoms, 85
   control, 86
Tomato spotted wilt virus
   control, 89
   symptoms, 87
Toxins, insect, 93
Trichodorus spp., 94
   primitivus, 101
   tobacco rattle virus vector, 81
Tubers
   abrasions, 14
   blotch, 84
   bruising, 16
   cracks, causes, 14
   description, 3
   greening and sunscald, 16
   indexing, viruses, 69
   respiration in storage, 5

rots, 66
   secondary, field and stored, 17
   seed treatment, 67
   wound healing, 3
Tylenchorhynchus spp., 94

Ulocladium
   atrum, symptoms, 46
   consortiale, symptoms, 46
Ulocladium blight, cause and symptoms, 46

Variation, *see Genetic abnormalities*
Verticillium
   albo-atrum, wilt caused by, 62
   dahliae, wilt caused by, 62
Verticillium wilt
   cause and symptoms, 62
   control, 63
Violet root rot
   cause and symptoms, 54
   control, 54
Virus diseases, 68

Wart
   cause and symptoms, 36
   control, 37
Whetzelina sclerotiorum, 49
White mold
   cause, 48
   control, 50
   symptoms, 49
Wildings, true and feathery, 7
Wilt, *see* Fusarium wilts and Verticillium wilt
Wind injury, 19
Witches' broom, 79
   control, 92
   symptoms, 92
Wound healing, in tubers, 3

Xanthomonas sp., in healthy tubers, 33
Xiphinema spp., 94
   americanum, tobacco ringspot virus vector,
      85
Xylaria sp., tuber rot from, 66

Yellow top, 91
Yield, worldwide, 1

Zinc deficiency, symptoms, 25